D0850054

Complex Variables and Transform Calculus

Complex Variables and Transform Calculus

by
M.Rahman

Technical University of Nova Scotia
Canada

Computational Mechanics Publications
Southampton UK and Boston USA

Matiur Rahman
Department of Applied Mathematics
Technical University of Nova Scotia
Halifax
Nova Scotia
Canada B3J 2X4

Published by

Computational Mechanics Publications
Ashurst Lodge, Ashurst, Southampton, SO40 7AA, UK
Tel: 44 (0)1703 293223 Fax: 44 (0)1703 292853
Email: cmp@cmp.co.uk
http://www.cmp.co.uk/

For USA, Canada and Mexico

Computational Mechanics Inc
25 Bridge Street, Billerica, MA 01821, USA
Tel: 508 667 5841 Fax: 508 667 7582
Email: cmina@ix.netcom.com

British Library Cataloguing-in-Publication Data

A Catalogue record for this book is available
from the British Library

ISBN 1 85312 491 5 Computational Mechanics Publications, Southampton

Library of Congress Catalog Card Number 96-086041   # 36864234

Printed and bound in Great Britain by Rexam Digital Imaging Limited.

Preface

The purpose of this book is to provide a modern treatment of complex variables and applications to the students of engineering and physical sciences. Although it is written primarily for the undergraduate students, certain parts of the text may be useful to graduate students and mathematical technologists working in industries. The applied complex variables are the most indispensable area of mathematics which have enormous applications in many problems arising in the industrial and physical sciences. This book will certainly **offshoot** the difficulties faced by the amateur scientists and engineers in their practical problems.

This book is comprised of eight chapters and the material in this book is arranged accordingly as follows:

The book is the outgrowth of my lecture notes for one semester course of engineering mathematics taught at the Technical University of Nova Scotia since 1980 to the undergraduate students in Electrical, Mechanical, Civil, Chemical, Industrial, Engineering Physics, Computer Science and Applied Mathematics. This book is, therefore, intended for a one-semester course at the undergraduate mathematics level basically needed by engineering students. However, Chapters 5, 6, 7 and 8 contain advanced materials and so it can also be used by the graduate students for their degree programs. The book will also be useful to the scientists and engineers who want to know the basic mathematical development of the theory of complex variables in solving their field problems.

Chapter 1 gives an outline of a complex number and its geometrical interpretation. This chapter also considers the analytic function of a complex variable and the Cauchy-Riemann equations. The integration in the complex plane is considered in Chapter 2. The Cauchy-Goursat theorem, the Cauchy integral formula are included in this chapter. Chapter 3 contains the Taylor series and the Laurent series expansion of a complex function. The theory of residues is considered in Chapter 4. The evaluation of real definite integrals and improper integrals with many illustrations has been considered in this chapter. A brief review of Fourier and Laplace transforms, the complex inversion integral of the Laplace transform theory with practical applications, stability criteria are discussed in Chapter 5. Chapter 6 gives an outline of the conformal mapping and the Schwarz-Christoffel transformations. Many applications of conformal mapping have been demonstrated in this chapter. Chapter 7 provides applications of complex variable theory to fluid mechanics, two dimensional potential theory, heat conduction, electrostatic and Dirichlet's problems. Chapter 8 is concerned with the analytic continuation of functions of complex variables.

Most of the basic results are stated as theorems omitting rigorous proofs. Usefulness of theorems are illustrated by examples and at the end of most of the sections exercises are given to practise the theory. Some problems in the exercises need considerable effort to solve. Therefore, references are provided aι the end of the book for further reading.

Some knowledge of vector calculus, solution techniques of ordinary and partial differential equations and elementary knowledge of integral transforms is assumed on the part of the students. For features special to the Laplace and Fourier transforms, readers are referred to the present author's *Applied Differential Equations for Scientists and Engineers* (Computational Mechanics Publications, Southampton, 1994). However, for matters concerned with applications of complex variables in depth, *Water Waves: Relating Modern Theory to Advanced Engineering Applications* (Oxford University Press, 1994) by the present author will be useful. For further insight into the complex variable applications to fluid mechanics problems, in particular, Lighthill's *An Informal Introduction to Theoretical Fluid Mechanics* (Oxford University Press, 1986) will be indispensable to the reader.

It is my very great pleasure to extend my gratitude to Sir James Lighthill, FRS for his continued interest and encouragement in the pursuit of my scientific endeavour. I am thankful to many of my colleagues and students who directly or indirectly helped me in preparation of the book. Mr Rezaul Abid deserves my appreciation for providing computing assistance and support at every step along the way. Special thanks are due to him for computer aided drawing and composition of all the figures of the book as well as his assistance in editing the entire manuscript. I would like to give special thanks to Mr Lance Sucharov and Professor Carlos Brebbia for their keen interest in the subject matter of this book. Financial support provided by the Natural Sciences and Engineering Research Council of Canada is gratefully acknowledged. I must express my gratitude to my family for their patience and understanding in the long preparation of this manuscript.

Halifax, Nova Scotia
February 1996   M. R.

Contents

Chapter 1

Complex numbers, analytic functions of a complex variable

1.1   Introduction

In many instances in the solution of practical problems the use of complex numbers plays a very important role. As for example, we use the theory of complex variables in solving differential equations. Its usefulness can also be seen in connection with the Fourier series and integrals, and complex inversion of Laplace transforms. Practical problems such as wave loads on offshore structures, ship movement in the oceans, wave scattering, spectral wave density etc. find its application tremendously useful. In this book we systematically develop the theory and provide important applications to the solution of these problems.

1.2   Algebraic properties, Cartesian and polar coordinates

Algebraic properties

It can be seen from the following, that there are equations which are not satisfied by any real numbers, for example

$$x^2 + 1 = 0, \qquad x^2 - 4x + 13 = 0$$

Roots of these equations are definitely not real and so this led to the introduction of complex numbers. In actual practice, the roots of the first equation are $\pm i$ and those of the second $2 \pm i3$ where i is defined as $i = \sqrt{-1}$.

By a complex number we mean a number of the form $z = x + iy$ where x and y are real numbers and i is the so-called imaginary unit whose definition is $i^2 = -1$.

The real numbers x and y are called the real and imaginary parts of z respectively; and are often denoted by $Re(z) = x$ and $Im(z) = y$. For example, $z = 4 - i3$, then $Re(z) = 4$ and $Im(z) = -3$.

Two complex numbers $z_1 = x_1 + iy_1$ and $z_2 = x_2 + iy_2$ are said to be equal if and only if the real and imaginary parts of the first are, respectively equal to the real and imaginary parts of the second. That means, if $z_1 = z_2$, then $x_1 = x_2$ and $y_1 = y_2$.

The operation of addition $(z_1 + z_2)$, subtraction $(z_1 - z_2)$ and multiplication $(z_1 z_2)$ are defined for the complex numbers $z_1 = x_1 + iy_1$ and $z_2 = x_2 + iy_2$ as follows:

$$
\begin{aligned}
z_1 + z_2 &= (x_1 + iy_1) + (x_2 + iy_2) \\
&= (x_1 + x_2) + i(y_1 + y_2) \tag{1.1} \\
z_1 - z_2 &= (x_1 + iy_1) - (x_2 + iy_2) \\
&= (x_1 - x_2) + i(y_1 - y_2) \tag{1.2} \\
z_1 z_2 &= (x_1 + iy_1)(x_2 + iy_2) \\
&= (x_1 x_2 - y_1 y_2) + i(x_1 y_2 + y_1 x_2) \tag{1.3}
\end{aligned}
$$

Also, if $z = x + iy$, then the negative of z is the complex number $-z = -x - iy$.

If two complex numbers differ only in the sign of their imaginary parts, then either one is said to be the conjugate of the other. The conjugate of a complex number is usually denoted by $\bar{z}$ or z^*. As for instance, if $z = x + iy$, then $\bar{z} = x - iy$.

Division of complex numbers is defined as the inverse of multiplication. In practice, the quotient of two complex numbers is usually found by multiplying both numerator and denominator by the conjugate of the denominator. As for example,

$$
\begin{aligned}
\frac{z_1}{z_2} = \frac{x_1 + iy_1}{x_2 + iy_2} &= \frac{(x_1 + iy_1)(x_2 - iy_2)}{(x_2 + iy_2)(x_2 - iy_2)} \\
&= \frac{(x_1 x_2 + y_1 y_2) + i(y_1 x_2 - x_1 y_2)}{x_2^2 + y_2^2} \\
&= z_3
\end{aligned}
$$

where

$$
\begin{aligned}
z_3 &= x_3 + iy_3 \\
&= \frac{x_1 x_2 + y_1 y_2}{x_2^2 + y_2^2} + i\frac{y_1 x_2 - x_1 y_2}{x_2^2 + y_2^2} \tag{1.4}
\end{aligned}
$$

provided $x_2^2 + y_2^2 \neq 0$.

One of the most striking properties of conjugate complex numbers is that

$$
z\bar{z} = (x + iy)(x - iy) = x^2 + y^2 = |z|^2 \tag{1.5}
$$

which is purely a real quantity. Also, $z + \bar{z} = (x + iy) + (x - iy) = 2x = 2Re(z)$
or

$$
Re(z) = \frac{1}{2}(z + \bar{z}) \tag{1.6}
$$

and $z - \bar{z} = (x + iy) - (x - iy) = 2iy = 2iIm(z)$
or

$$
Im(z) = \frac{1}{2i}(z - \bar{z}) \tag{1.7}
$$

Cartesian coordinates

A complex number $z = x + iy$ is represented with a point in the plane where Cartesian coordinates are x and y. At the same time the number z can be represented by the directed line segment or vector, from the origin to the point (x, y) as shown in Fig. 1.1.

The xy plane in which the complex numbers are represented in this manner is called the **complex plane**, or z plane or the **Argand diagram**, named after the French mathematician, Jean Robert Argand (1768-1822).

The sum and difference of two complex numbers $z_1 = x_1 + iy_1$ and $z_2 = x_2 + iy_2$ can be represented geometrically as shown in Fig. 1.2 (a) and (b). These figures are obtained by forming a parallelogram with the adjacent sides z_1 and z_2 with the proper directions.

Both the sum and difference of two complex numbers can be described in terms of the parallelogram having the given numbers for adjacent sides. The sum $z_1 + z_2$ is the diagonal which passes through the common origin of the two vectors and the difference $z_1 - z_2$ is the other diagonal, properly directed.

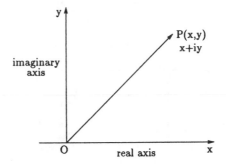

Fig. 1.1: The complex plane.

The complex conjugate, or simply the conjugate, of a complex number, $z = x + iy$ is defined as $\bar{z} = x - iy$ and is geometrically represented by Fig. 1.2 (c). The modulus or absolute value of a complex number $z = x + iy$ is defined as $|z| = \sqrt{x^2 + y^2}$ which is the distance between the point (x, y) and the origin. Thus the distance between two points representing the complex numbers z_1 and z_2 is given by

$$|z_1 - z_2| = \sqrt{(x_1 - x_2)^2 + (y_1 - y_2)^2}$$

The complex numbers corresponding to the points lying on the circle with centre (0, 2) and radius 4, for example, satisfy the equation $|z - 2i| = 4$. We refer to these points simply as circle.

The real number $|z|$, $Re(z)$ and $Im(z)$ are related by the relation

$$|z|^2 = [Re(z)]^2 + [Im(z)]^2 \tag{1.8}$$

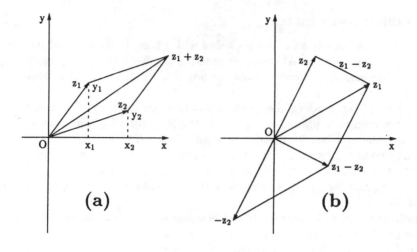

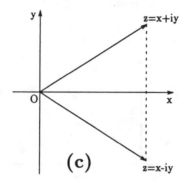

Fig. 1.2: The sum, difference and complex conjugates of complex numbers as shown in (a), (b) and (c).

as well as the inequalities

$$|z| \geq |Re(z)| \geq Re(z)$$
$$|z| \geq |Im(z)| \geq Im(z)$$

(1.9)

The following results are useful in taking the conjugate of a complex expression.

If $z_1 = x_1 + iy_1$ and $z_2 = x_2 + iy_2$, then

$$
\begin{aligned}
\overline{z_1 + z_2} &= \overline{(x_1 + x_2) + i(y_1 + y_2)} \\
&= (x_1 + x_2) - i(y_1 + y_2) \\
&= (x_1 - iy_1) + (x_2 - iy_2) \\
&= \bar{z}_1 + \bar{z}_2
\end{aligned}
$$

(1.10)

Similarly,

$$\overline{z_1 - z_2} = \bar{z}_1 - \bar{z}_2 \tag{1.11}$$

$$\overline{z_1 z_2} = \bar{z}_1 \bar{z}_2 \tag{1.12}$$

$$\overline{\left(\frac{z_1}{z_2}\right)} = \frac{\bar{z}_1}{\bar{z}_2}, \qquad z_2 \neq 0 \tag{1.13}$$

$$z\bar{z} = |z|^2 \tag{1.14}$$

The triangle inequality

The triangle inequality can be stated that if z_1 and z_2 are two complex numbers, then

$$|z_1 + z_2| \leq |z_1| + |z_2| \tag{1.15}$$

This property can be seen immediately from Fig. 1.3.

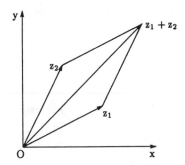

Fig. 1.3: Triangle inequality.

To see it mathematically, we write

$$\begin{aligned}
|z_1 + z_2|^2 &= (z_1 + z_2)(\overline{z_1 + z_2}) \\
&= (z_1 + z_2)(\bar{z}_1 + \bar{z}_2) \\
&= z_1\bar{z}_1 + z_1\bar{z}_2 + z_2\bar{z}_1 + z_2\bar{z}_2
\end{aligned}$$

But

$$\begin{aligned}
z_1\bar{z}_2 + \bar{z}_1 z_2 &= z_1\bar{z}_2 + \overline{z_1\bar{z}_2} \\
&= 2Re(z_1\bar{z}_2) \\
&\leq 2|z_1\bar{z}_2| \\
&= 2|z_1||z_2|
\end{aligned}$$

The right-hand side of the last expression is true because

$$|z_1 \bar{z}_2| = |z_1||\bar{z}_2|$$
$$= |z_1||z_2|$$

Thus

$$|z_1 + z_2|^2 \leq |z_1|^2 + 2|z_1||z_2| + |z_2|^2$$
$$= \{|z_1| + |z_2|\}^2$$

and hence $|z_1 + z_2| \leq |z_1| + |z_2|$ which is the required proof.

This property can be extended to any number of summands. Thus, if we write

$$|z_1 + z_2 + z_3| \leq |z_1 + z_2| + |z_3|$$
$$\leq |z_1| + |z_2| + |z_3|$$

which can be generalized to

$$\left| \sum_{k=1}^{n} z_k \right| \leq \sum_{k=1}^{n} |z_k| \qquad n = 1, 2, 3, \ldots \tag{1.16}$$

It can then be easily proved that $||z_1| - |z_2|| \leq |z_1 + z_2|$.

Polar coordinates

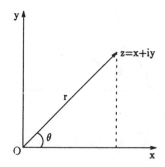

Fig. 1.4: Polar coordinates.

A complex number z can be represented in polar coordinates r and θ (Fig. 1.4), where

$$x = r \cos \theta$$
$$y = r \sin \theta$$

Thus

$$z = x + iy$$
$$= r \cos \theta + ir \sin \theta$$
$$= r(\cos \theta + i \sin \theta) \tag{1.17}$$

Therefore

$$r = \sqrt{x^2 + y^2} \tag{1.18}$$

$$\theta = \arctan\left(\frac{y}{x}\right) \tag{1.19}$$

Here the number r is the length of the vector representing z; that is $r = |z|$. The number θ is called the argument of z, and we write $\theta = \arg z$. The principal value of $\arg z$, denoted by $\text{Arg } z$ is defined as that unique value of $\arg z$ such that $-\pi < \arg z < \pi$. If $z = 0$, then θ is undefined which is obvious from (1.19). The polar form of $z - z_0$ (Fig. 1.5) when $z \neq z_0$ is

$$z - z_0 = \rho(\cos\phi + i\sin\phi)$$

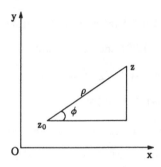

Fig. 1.5: Polar form of $(z - z_0)$.

The product and division of two complex numbers z_1 and z_2 can be obtained from the polar representation of a complex number (see Fig. 1.6).
If we have two complex numbers given in polar form, their product can be written

$$\begin{aligned}
z_1 z_2 &= (r_1\cos\theta_1 + ir_1\sin\theta_1)(r_2\cos\theta_2 + ir_2\sin\theta_2) \\
&= r_1 r_2[\{\cos\theta_1\cos\theta_2 - \sin\theta_1\sin\theta_2\} \\
&\quad + i\{\cos\theta_1\sin\theta_2 + \sin\theta_1\cos\theta_2\}] \\
&= r_1 r_2[\cos(\theta_1 + \theta_2) + i\sin(\theta_1 + \theta_2)]
\end{aligned} \tag{1.20}$$

and then the quotient can be written

$$\begin{aligned}
\frac{z_1}{z_2} &= \frac{r_1(\cos\theta_1 + i\sin\theta_1)}{r_2(\cos\theta_2 + i\sin\theta_2)} \\
&= \frac{r_1}{r_2}\frac{(\cos\theta_1 + i\sin\theta_1)(\cos\theta_2 - i\sin\theta_2)}{(\cos\theta_2 + i\sin\theta_2)(\cos\theta_2 - i\sin\theta_2)} \\
&= \frac{r_1}{r_2}\frac{(\cos\theta_1\cos\theta_2 + \sin\theta_1\sin\theta_2) + i(\sin\theta_1\cos\theta_2 - \cos\theta_1\sin\theta_2)}{\cos^2\theta_2 + \sin^2\theta_2} \\
&= \frac{r_1}{r_2}[\cos(\theta_1 - \theta_2) + i\sin(\theta_1 - \theta_2)], \qquad r_2 \neq 0
\end{aligned} \tag{1.21}$$

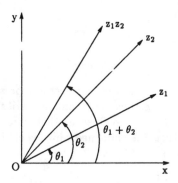

Fig. 1.6: Multiplication of two complex numbers.

Thus it can be seen from (1.20) and (1.21) that

$$\arg(z_1 z_2) \quad = \quad \arg z_1 + \arg z_2 \tag{1.22}$$

$$\arg\left(\frac{z_1}{z_2}\right) \quad = \quad \arg z_1 - \arg z_2 \tag{1.23}$$

These ideas can be extended to products of more than two factors and can be written

$$z_1 z_2 \cdots z_n \quad = \quad r_1 r_2 \cdots r_n [\cos(\theta_1 + \theta_2 + \cdots + \theta_n) \\ + i\sin(\theta_1 + \theta_2 + \cdots + \theta_n)]$$

In particular, if all the zs are the same, we have that

$$z^n = r^n[\cos n\theta + i\sin n\theta] \tag{1.24}$$

when $r = 1$, then

$$z^n = \cos n\theta + i\sin n\theta$$

or $\qquad (\cos\theta + i\sin\theta)^n = \cos n\theta + i\sin n\theta \tag{1.25}$

which is known as **Demoivre's theorem** named after the French mathematician, Abraham Demoivre (1667-1754).

The expansion of $e^{i\theta}$ by Taylor series suggests that it is often convenient to let $e^{i\theta}$ denote the expression $\cos\theta + i\sin\theta$:

$$e^{i\theta} = \cos\theta + i\sin\theta \tag{1.26}$$

This is known as **Euler's formula** named after the Swiss mathematician, Leonhard Euler (1707-1783). Thus a complex number z can be written as

$$\begin{aligned} z \quad &= \quad x + iy \\ &= \quad r(\cos\theta + i\sin\theta) \\ &= \quad re^{i\theta} \end{aligned}$$

and hence

$$z^n = r^n e^{ni\theta} \tag{1.27}$$

The formula (1.27) is helpful, in computing roots of non-zero complex numbers. As for example, to find the n roots of unity, we start with the equation

$$
\begin{aligned}
z^n &= 1 \\
&= \cos 2\pi + i \sin 2\pi \\
&= \cos 2k\pi + i \sin 2k\pi \\
&= e^{i2k\pi}
\end{aligned}
$$

Therefore,

$$z = e^{i(2k\pi/n)} \qquad k = 0, 1, 2, ..., (n-1) \tag{1.28}$$

Thus we have n distinct roots of unity.

If we want to find the cube root of unity, we have

$$
\begin{aligned}
z^3 &= 1 \\
&= \cos 2k\pi + i \sin 2k\pi \\
&= e^{i(2k\pi)}
\end{aligned}
$$

Therefore,

$$z = e^{i(2k\pi/3)} \qquad k = 0, 1, 2$$

Thus the 3 roots of unity can be obtained from Fig. 1.7.

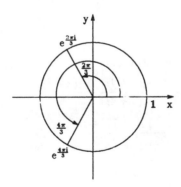

Fig. 1.7: Unit circle showing the roots of unity.

The above idea can be easily extended to finding the nth roots of any non-zero complex number

$$
\begin{aligned}
z &= r(\cos\theta + i\sin\theta) \\
&= r[\cos(\theta + 2\pi) + i\sin(\theta + 2\pi)]
\end{aligned}
$$

Then in general

$$z = r[\cos(\theta + 2k\pi) + i\sin(\theta + 2k\pi)]$$

where k is any integer multiple of 2π. Therefore, $z = re^{i(\theta + 2k\pi)}$.

Thus

$$
\begin{aligned}
w &= z^{\frac{1}{n}} \\
&= r^{\frac{1}{n}}e^{i\left(\frac{\theta+2k\pi}{n}\right)} \\
&= r^{\frac{1}{n}}\left[\cos\left(\frac{\theta+2k\pi}{n}\right) + i\sin\left(\frac{\theta+2k\pi}{n}\right)\right]
\end{aligned}
\tag{1.29}
$$

where $k = 0, 1, 2, \dots (n-1)$.

In the complex plane these are represented by radii of the circle with centre at the origin and radii $r^{\frac{1}{n}}$ spaced at equal angular intervals of $2\pi/n$, beginning with the radius whose angle is θ/n.

Similarly we can define for general rational function,

$$
\begin{aligned}
z^{\frac{p}{q}} &= \left[re^{i(\theta+2k\pi)}\right]^{\frac{p}{q}} \\
&= r^{\frac{p}{q}}e^{i\frac{p}{q}(\theta+2k\pi)} \\
&= r^{\frac{p}{q}}\left[\cos\frac{p}{q}(\theta+2k\pi) + i\sin\frac{p}{q}(\theta+2k\pi)\right] \quad k = 0,1,2,3,\dots(q-1)
\end{aligned}
\tag{1.30}
$$

Example 1

Verify

$$\frac{1+2i}{3-4i} + \frac{2-i}{5i} = -\frac{2}{5}$$

Solution

Considering the left hand side we obtain

$$
\begin{aligned}
\frac{1+2i}{3-4i} + \frac{2-i}{5i} &= \frac{(1+2i)(3+4i)}{(3-4i)(3+4i)} - \frac{i(2-i)}{5} \\
&= \frac{-5+10i}{25} - \frac{10i+5}{25} \\
&= -\frac{10}{25} \\
&= -\frac{2}{5}
\end{aligned}
$$

Example 2

Show that each of the two numbers $z = 1 \pm i$ satisfies the equation

$$z^2 - 2z + 2 = 0$$

Solution

Solving the given quadratic equation for z we have

$$z^2 - 2z + 2 = 0$$
$$z = \frac{2 \pm \sqrt{4-8}}{2}$$
$$= 1 \pm i$$

Hence as asserted.

Example 3

Show that

$$\frac{(2+i)^2}{3-4i} = 1$$

Solution

$$\frac{(2+i)^2}{3-4i} = \frac{3+4i}{3-4i} = \frac{3-4i}{3-4i} = 1$$

Hence as asserted.

Example 4

Use the polar form to show that

$$i(1 - i\sqrt{3})(\sqrt{3}+i) = 2 + i\,2\sqrt{3}$$

Solution

$$i = \cos\frac{\pi}{2} + i\sin\frac{\pi}{2} = e^{i\pi/2}$$
$$1 - \sqrt{3}i = 2e^{-i\,\tan^{-1}\sqrt{3}} = 2e^{-i\,\pi/3}$$
$$\sqrt{3} + i = 2e^{i\,\tan^{-1}1/\sqrt{3}} = 2e^{i\,\pi/6}$$

Therefore,

$$i(1-\sqrt{3})(\sqrt{3}+i) = (e^{i\pi/2})(2e^{-i\pi/3})(2e^{i\pi/6})$$
$$= 4e^{i(\pi/2-\pi/3+\pi/6)}$$
$$= 4e^{i\pi/3}$$
$$= 4\left[\cos\frac{\pi}{3} + i\sin\frac{\pi}{3}\right]$$
$$= 4\left[\frac{1}{2} + i\frac{\sqrt{3}}{2}\right]$$
$$= 2 + i\,2\sqrt{3}$$

Hence this is the required proof.

Exercises

Reduce each of the following operations to the form $a + ib$ where a and b are both real numbers:

1. $(3+i)(1+2i)$

2. $\dfrac{1-3i}{2+i}$

3. $(1-2i)(2+3i)^2$

4. $3\left(\dfrac{1-i}{1+i}\right)^2 - 2\left(\dfrac{1+i}{1-i}\right)^3$

5. $\dfrac{(1+i)^3}{(2+i)(1+2i)}$

6. $\dfrac{i^3 - i^8 + i^{15}}{1 - i^4 + i^9 - i^{14}}$

Solve each of the following equations:

7. $z^2 - z + 1 = 0$

8. $z\bar{z} - z + \bar{z} = 0$

9. $z^3 = i$

10. $z^2 + 2\bar{z} = 1 - 4i$

11. $(x^2 y - 2) + (x + 2xy - 5)i = 0$ for all x and y

12. Prove that

    (a) $Re(z) = \dfrac{z + \bar{z}}{2}$

    (b) $Im(z) = \dfrac{z - \bar{z}}{2i}$

13. If $f(z)$ is a polynomial in z with real coefficients and $f(z + 2i) = 5 - iz$, then find the value of $f(z - 2i)$.

14. Describe and sketch the locus represented by each of the following:

    (a) $|z - i| = 1$

    (b) $|z + 4i| + |z - i| = 4$

    (c) $|z - 2| - |z + 2| = 4$

    (d) $z(1 + \bar{z}) = 2$

    (e) $I_m(z^2) = 2$

15. Under what conditions, if any, will the equation

$$a\,z^2 + 2b\,z\bar{z} + c\,\bar{z}^2 = d,$$

   a, b, c, and d real, represent

   (a) a hyperbola?
   (b) a parabola?
   (c) an ellipse?
   (d) a circle?

16. If $b\bar{b} > (a + \bar{a})(c + \bar{c})$, show that the equation

$$(a + \bar{a})z\bar{z} + bz + \bar{b}\bar{z} + (c + \bar{c}) = 0$$

   represents a real circle, and find its centre and radius.

17. Express the complex number $(3 + \sqrt{2}i)$ in polar form and find its distinct fifth roots.

18. Find all the distinct square roots of -1.

19. Find all the distinct square roots of 32.

20. Find all the distinct cube roots of i.

21. Find all the distinct cube roots of $1 + i$.

22. Find all the distinct values of $(1+i)^{3/2}$. What are the distinct values of $(1+i)^{2/3}$?

23. Using Demoivre's theorem, express $\cos 3\theta$ and $\sin 3\theta$ in terms of powers of $\cos \theta$ and $\sin \theta$.

1.3 Functions of a complex variable, limits and continuity

We shall introduce in this section some basic concepts of functions of a complex variable; limits and continuity. These concepts can be seen to be similar to those of a real variable in calculus.

If $z = x + iy$ and $w = u + iv$ are two complex variables, and if for each value of z in some domains of the complex plane, one or more values of w are defined, then w is said to be a function of z, and we write $w = f(z)$ which can also be written as

$$w = u + iv = f(z) \tag{1.31}$$

where the variables u, v, x, y are all real, and $u(x,y)$ and $v(x,y)$ are two real valued functions of x and y.

Consider the function $w = z^2$. By substituting the general polar value of $z = re^{i(\theta+2n\pi)}$ into this function we obtain $w = r^2 e^{i(2\theta+4n\pi)}$ which simply is $w = r^2 e^{i2\theta}$ for all integral values of n ($n = 0, 1, 2, 3...$). Hence to every value of z there corresponds one and only one value of w, so that the latter is said to be a single valued function of z. Moreover for all finite values of z, w is defined uniquely. This functional relationship between w and z could be plotted graphically if they were both real variables. However, with complex numbers this is impossible, since there are four variables. The concept of geometrical representation of the functional relationship between w and z has been discussed extensively in Chapter 6 in connection with the conformal mapping. It is worth mentioning here, however, that the absolute value of w can be very easily plotted as the modular surface considering three dimensional coordinate system (see McLachlan (1963)).

In the following we demonstrate two instances how to obtain relationship between u, v, x, and y. First one deals with how to express $u(x,y)$ and $v(x,y)$ as functions of x and y; whereas the second expresses the relationship between w and z in complex variables.

As for example, we can write the function $w = f(z) = z^2 + 2z$ for all z. Here

$$
\begin{aligned}
u + iv &= (x+iy)^2 + 2(x+iy) \\
&= (x^2 - y^2 + 2x) + 2i(xy+y)
\end{aligned}
$$

so that on taking real and imaginary parts, we have

$$
\begin{aligned}
u(x,y) &= x^2 - y^2 + 2x \\
v(x,y) &= 2(xy + y)
\end{aligned}
$$

Also from such a pair of real functions, we can obtain a complex function of z. For example:

Let us consider $u(x,y) = x/(x^2+y^2)$ and $v(x,y) = -y/(x^2+y^2)$ then

$$
\begin{aligned}
w &= u + iv \\
&= \frac{x}{x^2+y^2} - \frac{iy}{x^2+y^2} \\
&= \frac{x-iy}{x^2+y^2} = \frac{1}{x+iy} = \frac{1}{z}
\end{aligned}
$$

Limits

If a function $f(z)$ is defined at all points in some neighbourhood of z_0, except possibly at the point z_0 itself, then the statement that the limit of f as z approaches z_0 is a number w_0, such that

$$
\lim_{z \to z_0} f(z) = w_0 \tag{1.32}
$$

The meaning of the above statement is that the point $w = f(z)$ can be made arbitrarily close to w_0 if we choose a point z which is close enough to z_0 but distinct from it. Another way of putting the statement (1.32) is that for each positive number ε there is a positive number δ such

$$|f(z) - w_0| < \varepsilon \qquad \text{whenever } |z - z_0| < \delta \qquad (1.33)$$

Continuity

A function $f(z)$ is said to be continuous at a point z_0 provided all three of the following conditions are statisfied:

(i) $\lim_{z \to z_0} f(z)$ exists

(ii) $f(z_0)$ exists

(iii) $\lim_{z \to z_0} f(z) = f(z_0)$

In other words, for a function to be continuous at a point z_0, the function must have both a value at that point and a limit as z approaches that point and the two must be equal.

Another way we can say that $f(z)$ is said to be continuous at z_0, if for each positive number ε there exists a positive number δ such that

$$|f(z) - f(z_0)| < \varepsilon \qquad \text{whenever } |z - z_0| < \delta \qquad (1.34)$$

A function of a complex variable is said to be continuous in a region R if it is continuous at each point in R.

Example 1

Show that

$$\lim_{z \to 2i} \frac{z^2 + 4}{z - 2i} = 4i$$

Solution

We have

$$\lim_{z \to 2i} \frac{z^2 + 4}{z - 2i} = \lim_{z \to 2i} \frac{(z + 2i)(z - 2i)}{(z - 2i)} \qquad \left(\frac{0}{0} \text{ form}\right)$$
$$= 2i + 2i$$
$$= 4i$$

Hence as asserted.

Example 2

Find

$$\lim_{z \to 1} \frac{2z}{z + i}$$

Solution

$$\lim_{z \to 1} \frac{2z}{z + i} = \frac{2}{1 + i}$$
$$= \frac{2(1 - i)}{(1 + i)(1 - i)} = \frac{2(1 - i)}{2}$$
$$= 1 - i$$

Example 3

Find

$$\lim_{z \to \pi i} \frac{\sin z + z}{2 + e^z}$$

Solution

$$\lim_{z \to \pi i} \frac{\sin z + z}{2 + e^z} = \lim_{z \to \pi i} \frac{\sin(\pi i) + \pi i}{2 + e^{\pi i}}, \qquad (\sin \pi i = i \sinh \pi)$$
$$= \frac{i \sinh \pi + \pi i}{2 + (\cos \pi + i \sin \pi)}$$
$$= \frac{i(\pi + \sinh \pi)}{2 - 1}$$
$$= i(\pi + \sinh \pi)$$

Example 4

Determine at what points, if any, the following function fails to be continuous and explain why:

$$f(z) = \frac{z^2 - 1}{z - 1}$$

Solution

In this function, $z = 1$ is the point where $f(z)$ is undefined.

Now if $f(1) = 2$, then $f(z)$ is continuous at $z = 1$ because

$$f(z) = \frac{z^2 - 1}{z - 1}$$

$$\lim_{z \to 1} f(z) = \lim_{z \to 1} \frac{z^2 - 1}{z - 1}$$

$$= \lim_{z \to 1} \frac{(z + 1)(z - 1)}{(z - 1)}$$

$$= 2$$

$$= f(1)$$

Exercises

1. For each of the following functions describe the domain of definition:

   (a) $f(z) = \dfrac{1}{z - 1}$

   (b) $f(z) = \dfrac{1}{z^2 + 1}$

   (c) $f(z) = \arg\left(\dfrac{1}{z}\right)$

   (d) $f(z) = z + \dfrac{1}{z}$

2. If $f(z) = xy + i(x^2 + y^2)$, then find $f(1 + 2i)$.

3. If $f(z) = z^2 + \bar{z} + Re(z\bar{z})$, what is $f(2 + i)$?

4. Express $x^2 + iy^2$ in terms of z and $\bar{z}$.

5. Express $x^2 - y^2 - 2y + i(2x - 2xy)$ in terms of z and $\bar{z}$.

6. Sketch each of the following domains:

   (a) $Re(z) > 0$
   (b) $Im(z) > 0$
   (c) $0 \le Re(z) \le 1$
   (d) $|z - 2| > 2$
   (e) $1 < |z| < 2$
   (f) $0 \le Im(z) < Re(z)$

7. Show that

   (a) $\displaystyle \lim_{z \to i} \frac{z^2 + 1}{z - i} = 2i$

(b) $\lim\limits_{z \to 1} \dfrac{2z}{z+i} = 1 - i$

(c) $\lim\limits_{z \to 1-i} [x + i(2x + y)] = 1 + i$

(d) $\lim\limits_{z \to z_0} \dfrac{1}{z^n} = \dfrac{1}{z_0^n} \quad (z_0 \neq 0)$

(e) $\lim\limits_{z \to i} \dfrac{i\,z^3 - 1}{z+i} = 0$

8. Using a definition of limit involving the point at infinity, prove that

(a) $\lim\limits_{z \to \infty} \dfrac{z}{z^2 + 1} = 0$

(b) $\lim\limits_{z \to 0} \dfrac{1}{z} = \infty$

(c) $\lim\limits_{z \to 1} \dfrac{2}{(z-1)^2} = \infty$

(d) $\lim\limits_{z \to \infty} 2z^2 = \infty$

9. Explain why $\lim\limits_{z \to 0} \left[\dfrac{xy}{x^2 + y^2} \right]$ does not exist.

10. Explain why $\lim\limits_{z \to 0} \left[\dfrac{x^2 y}{x^4 + y^2} \right]$ does not exist.

11. Determine at what points, if any, each of the following functions fails to be continuous:

(a) $f(z) = \dfrac{z}{z^2 - 1}$

(b) $f(z) = \dfrac{z^2 + 1}{z - i}$

(c) $f(z) = \begin{cases} (z^2 + 2iz - 1)/(z+i) & z \neq -i \\ i & z = -i \end{cases}$

(d) $f(z) = \begin{cases} z^2 + 2iz + 2 & z \neq i \\ i & z = i \end{cases}$

12. Prove that $\dfrac{d}{dz}(\bar{z}z^2)$ does not exist anywhere.

13. Determine whether $|z|^2$ has a derivative somewhere.

1.4 Analytic functions, Cauchy-Riemann equations and Laplace's equation

The derivatives of a function of a complex variable $w = f(z)$ is defined to be

$$\frac{dw}{dz} = w' = f'(z)$$

$$= \lim_{\Delta z \to 0} \left\{ \frac{f(z + \Delta z) - f(z)}{\Delta z} \right\} \tag{1.35}$$

provided the limit here exists. The function f is said to be differentiable at z when its derivative exists at z. In other words, (1.35) can be written as follows:

$f(z)$ is said to be differentiable at z provided that for every ε there exists a number δ such that

$$\left| \frac{f(z + \Delta z) - f(z)}{\Delta z} - f'(z) \right| < \varepsilon \tag{1.36}$$

whenever $0 < \Delta z < \delta$.

This definition of derivative is formally identical in form to that of the derivative of a real valued function of a real variable. In fact, it is clear that formulas for the differentiation of functions of a real variable are equally applicable to the case of functions of a complex variable with proper definition. Thus it is easy to verify that such familiar formulas as

$$\frac{dc}{dz} = 0$$

$$\frac{dz}{dz} = 1$$

$$\frac{d}{dz}(cw) = cw'$$

$$\frac{d}{dz}(w_1 \pm w_2) = \frac{dw_1}{dz} \pm \frac{dw_2}{dz}$$

$$\frac{d}{dz}(w_1 w_2) = w_1 \frac{dw_2}{dz} + w_2 \frac{dw_1}{dz}$$

$$\frac{d}{dz}\left(\frac{w_1}{w_2}\right) = \frac{w_2 \left(\frac{dw_1}{dz}\right) - w_1 \left(\frac{dw_2}{dz}\right)}{w_2^2} \qquad w_2 \neq 0 \text{ for every } z$$

$$\frac{d}{dz}(w^n) = nw^{n-1}\frac{dw}{dz}$$

are valid when w_1, w_2, and w are differentiable functions of a complex variable z and c is a constant complex number.

The Cauchy-Riemann equations

Suppose that a function $f(z)$ is defined on a neighbourhood of a point z by the equation

$$f(z) = u(x, y) + iv(x, y)$$

and given that $f'(z)$ exists at z. Then, we have

$$\Delta f = f(z + \Delta z) - f(z)$$

also $$\Delta f = \Delta u(x, y) + i\,\Delta v(x, y)$$

where

$$\Delta u = u(x + \Delta x, y + \Delta y) - u(x, y)$$
$$\Delta v = v(x + \Delta x, y + \Delta y) - v(x, y)$$
$$\Delta z = \Delta x + i\,\Delta y$$

Now

$$\lim_{\Delta z \to 0}\left(\frac{\Delta f}{\Delta z}\right) = \lim_{\Delta z \to 0}\left\{\frac{f(z + \Delta z) - f(z)}{\Delta z}\right\}$$

$$= \lim_{\Delta x, \Delta y \to 0}\left\{\frac{\Delta u + i\,\Delta v}{\Delta x + i\,\Delta y}\right\} \tag{1.37}$$

Equation (1.37) can be written

$$\lim_{\Delta x, \Delta y \to 0}\left\{\frac{\Delta u + i\,\Delta v}{\Delta x + i\,\Delta y}\right\} = a + i\,b \tag{1.38}$$

where

$$a = \lim_{\Delta x, \Delta y \to 0} Re\left\{\frac{\Delta u + i\,\Delta v}{\Delta x + i\,\Delta y}\right\}$$

$$b = \lim_{\Delta x, \Delta y \to 0} Im\left\{\frac{\Delta u + i\,\Delta v}{\Delta x + i\,\Delta y}\right\}$$

In Fig. 1.8, it is clear that Δz can approach zero in infinitely many ways. Now if Δz is real, i.e. if $\Delta y = 0$, we obtain from (1.38) that

$$f'(z) = \frac{\partial f}{\partial z} = \lim_{\Delta x \to 0}\left\{\frac{\Delta u + i\,\Delta v}{\Delta x}\right\} = \frac{\partial u}{\partial x} + i\frac{\partial v}{\partial x} = a + i\,b \tag{1.39}$$

Therefore equating real and imaginary parts:

$$\frac{\partial u}{\partial x} = a \tag{1.40}$$

$$\frac{\partial v}{\partial x} = b$$

On the other hand, if Δz is imaginary, i.e., $\Delta x = 0$, we obtain from (1.38) that

$$f'(z) = \lim_{\Delta y \to 0}\left\{\frac{\Delta u + i\,\Delta v}{i\Delta y}\right\} = -i\frac{\partial u}{\partial y} + \frac{\partial v}{\partial y} = a + i\,b \tag{1.41}$$

Therefore equating real and imaginary parts:

$$\frac{\partial v}{\partial y} = a \tag{1.42}$$

$$\frac{\partial u}{\partial y} = -b$$

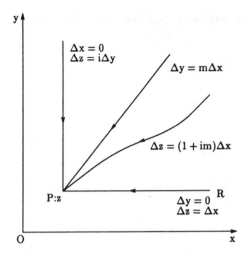

Fig. 1.8: Various ways in which Δz can approach zero.

Thus comparing (1.40) and (1.42), we have

$$\frac{\partial u}{\partial x} = \frac{\partial v}{\partial y} \qquad (1.43)$$

$$\frac{\partial u}{\partial y} = -\frac{\partial v}{\partial x}$$

These two extremely important conditions, which are known as the Cauchy-Riemann equations, are so named in honour of the French mathematician, A.L. Cauchy (1789-1857), who discovered them, and in honour of the German mathematician, G.F.B. Riemann (1826-1866), who made them fundamental in his development of the theory of functions of a complex variable.

Now we can define the following. If $w = f(z)$ possesses a derivative at z and at every point in some neighbourhood of z, $f(z)$ is said to be analytic at z and z is called the *regular* point of the function. If $f(z)$ is not analytic at z, but if every neighbourhood of z contains points at which $f(z)$ is analytic, then z is called a singular point of $f(z)$.

Theorem 1.1

If u and v are real single-valued functions of x and y which, with u_x, u_y, v_x and v_y are continuous throughout a region R, then Cauchy-Riemann conditions

$$\frac{\partial u}{\partial x} = \frac{\partial v}{\partial y}$$

$$\frac{\partial u}{\partial y} = -\frac{\partial v}{\partial x}$$

are both *necessary* and *sufficient* conditions that $f(z) = u(x,y) + i\,v(x,y)$ be analytic in R.

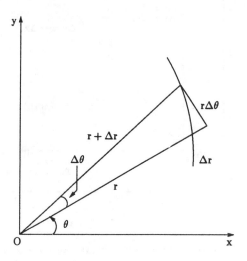

Fig. 1.9: Polar representation.

Thus the derivative of $f(z)$ is given either by

$$f'(z) = \frac{\partial u}{\partial x} + i\,\frac{\partial v}{\partial x}$$

or   (1.44)

$$f'(z) = \frac{\partial v}{\partial y} - i\,\frac{\partial u}{\partial y}$$

The Cauchy-Riemann equations in polar form can be obtained by means of the coordinate transformation (Fig. 1.9) $x = r\cos\theta$ and $y = r\sin\theta$ such that $r = \sqrt{(x^2 + y^2)}$ and $\theta = \arctan(y/x)$.

With this transformation,

$$\frac{\partial u}{\partial x} = \frac{\partial u}{\partial r}\frac{\partial r}{\partial x} + \frac{\partial u}{\partial \theta}\frac{\partial \theta}{\partial x}$$

$$\frac{\partial v}{\partial y} = \frac{\partial v}{\partial r}\frac{\partial r}{\partial y} + \frac{\partial v}{\partial \theta}\frac{\partial \theta}{\partial y}$$

Using the fact that

$$\frac{\partial r}{\partial x} = \frac{x}{r} = \cos\theta$$

$$\frac{\partial r}{\partial y} = \frac{y}{r} = \sin\theta$$

$$\frac{\partial \theta}{\partial x} = -\frac{y}{r^2} = -\frac{\sin\theta}{r}$$

$$\frac{\partial \theta}{\partial y} = \frac{x}{r^2} = \frac{\cos\theta}{r}$$

the Cauchy-Riemann equations in polar form can be written after equating $\cos\theta$ and $\sin\theta$ terms from the relation $\partial u/\partial x = \partial v/\partial y$:

$$\frac{\partial u}{\partial r} = \frac{\partial v}{r\partial\theta} \qquad (1.45)$$

$$\frac{\partial u}{r\partial\theta} = -\frac{\partial v}{\partial r}$$

The explicit derivation of these formulae is left as an exercise for the reader.

Thus the Cauchy-Riemann theorem can be explicitly stated as follows:

Necessary conditions

If $f(z) = u(x, y) + i\, v(x, y)$ and its derivative $f'(z)$ exist, then,

$$\frac{\partial u}{\partial x} = \frac{\partial v}{\partial y}$$

$$\frac{\partial u}{\partial y} = -\frac{\partial v}{\partial x}$$

Sufficient conditions

If

$$\frac{\partial u}{\partial x} = \frac{\partial v}{\partial y}$$

$$\frac{\partial u}{\partial y} = -\frac{\partial v}{\partial x}$$

then $f'(z)$ exists where $f(z) = u(x, y) + i\, v(x, y)$.

Laplace's equation

If both the real part and the imaginary part of an analytic function $f(z)$ have continuous second partial derivatives, then they satisfy Laplace's equation,

$$\frac{\partial^2\phi}{\partial x^2} + \frac{\partial^2\phi}{\partial y^2} = 0$$

Proof

This can be easily proved using the Cauchy-Riemann conditions of an analytic function $f(z) = u(x, y) + i\, v(x, y)$. The Cauchy-Riemann conditions are:

$$\frac{\partial u}{\partial x} = \frac{\partial v}{\partial y}$$

$$\frac{\partial u}{\partial y} = -\frac{\partial v}{\partial x}$$

If we differentiate the first of these with respect to x and the second with respect to y, and add the result, we obtain

$$\frac{\partial^2 u}{\partial x^2} = \frac{\partial^2 v}{\partial x\, \partial y}$$

$$\frac{\partial^2 u}{\partial y^2} = -\frac{\partial^2 v}{\partial y\, \partial x}$$

So

$$\frac{\partial^2 u}{\partial x^2} + \frac{\partial^2 u}{\partial y^2} = 0 \qquad (1.46)$$

Similarly, if we differentiate the first with respect to y and the second with respect to x, and subtract the result, we obtain

$$\frac{\partial^2 v}{\partial x^2} + \frac{\partial^2 v}{\partial y^2} = 0 \qquad (1.47)$$

A function which possesses continuous second partial derivatives and satisfies Laplace's equation is usually called a **harmonic function**. Two harmonic functions u and v so related that $u + i\, v$ is an analytic function are called **conjugate harmonic** functions. The word conjugate here is not the same as conjugate of a complex number z.

Orthogonality

If $w = f(z) = u(x,y) + iv(x,y)$ is an analytic function of z, then the curves of the family $u(x,y) = c_1$ are orthogonal trajectories of the curves of the family $v(x,y) = c_2$ and vice versa.

Proof

To prove this, we compute the slope of the general curve of each family. Thus if $u(x,y) = c_1$, then $du(x,y) = 0$ and so $(\partial u/\partial x)dx + (\partial u/\partial y)dy = 0$.

Therefore the slope of the first family of curves is

$$\left(\frac{dy}{dx}\right)_1 = -\frac{\partial u/\partial x}{\partial u/\partial y} \qquad (1.48)$$

Also, we have from the second family of curves $v(x,y) = c_2$, then $dv(x,y) = 0$ and hence $(\partial v/\partial x)dx + (\partial v/\partial y)dy = 0$.

Thus we have the slope of the second family of curves as

$$\left(\frac{dy}{dx}\right)_2 = -\left(\frac{\partial v/\partial x}{\partial v/\partial y}\right) \qquad (1.49)$$

Then the product of the two slopes after using the Cauchy-Riemann conditions becomes

$$\left(\frac{dy}{dx}\right)_1 \left(\frac{dy}{dx}\right)_2 = \left(-\frac{\partial u/\partial x}{\partial u/\partial y}\right) \left(-\frac{\partial v/\partial x}{\partial v/\partial y}\right)$$

$$= \left\{-\frac{\partial u/\partial x}{\partial u/\partial y}\right\} \left\{\frac{\partial u/\partial y}{\partial u/\partial x}\right\}$$

$$= -1$$

Since the product is negative unity, therefore the families of curves

$$u(x,y) = c_1 \qquad \text{and} \qquad v(x,y) = c_2$$

are orthogonal to each other as shown in Fig. 1.10.

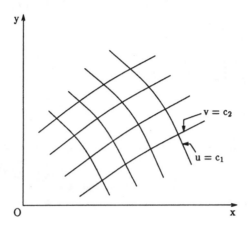

Fig. 1.10: Orthogonality of $u(x,y) = c_1$ and $v(x,y) = c_2$.

Example 1

Show that the derivative of the function $f(z) = \bar{z}$ does not exist.

Solution

Given that

$$f(z) = \bar{z} = x - i\,y$$

then by definition,

$$
\begin{aligned}
f'(z) &= \lim_{\Delta z \to 0} \left\{ \frac{f(z + \Delta z) - f(z)}{\Delta z} \right\} \\
&= \lim_{\Delta x, \Delta y \to 0} \left\{ \frac{[(x + \Delta x) - i(y + \Delta y)] - [x - i\,y]}{\Delta x + i\,\Delta y} \right\} \\
&= \lim_{\Delta x, \Delta y \to 0} \left\{ \frac{\Delta x - i\,\Delta y}{\Delta x + i\,\Delta y} \right\}
\end{aligned}
$$

Now let Δz be real, such that $\Delta y = 0$ and hence

$$f'(z) = \lim_{\Delta x \to 0} \frac{\Delta x}{\Delta x} = 1$$

Also let Δz be imaginary, such that $\Delta x = 0$ and hence

$$f'(z) = \lim_{\Delta y \to 0} \left\{ \frac{-i\,\Delta y}{i\,\Delta y} \right\} = -1$$

More generally, if we let $\Delta z \to 0$ in such a way that $\Delta y = m\,\Delta x$, we have

$$
\begin{aligned}
f'(z) &= \lim_{\Delta x, \Delta y \to 0} \left(\frac{\Delta x - i\,\Delta y}{\Delta x + i\,\Delta y} \right) \\
&= \lim_{\Delta x \to 0} \left(\frac{\Delta x - im\,\Delta x}{\Delta x + im\,\Delta x} \right) \\
&= \frac{1 - im}{1 + im} \\
&= \frac{(1 - m^2) - 2im}{1 + m^2}, \qquad m \text{ arbitrary}
\end{aligned}
$$

Thus there are infinitely many values of the derivative of $f(z)$ existing which clearly indicates that $f(z) = \bar{z} = x - i\,y$ has no derivative at all.

Also we can prove that Cauchy-Riemann conditions are not satisfied in this problem. Because

$$
\begin{aligned}
u &= x, \qquad v = -y \\
\frac{\partial u}{\partial x} &= 1, \qquad \frac{\partial v}{\partial x} = 0 \\
\frac{\partial u}{\partial y} &= 0, \qquad \frac{\partial v}{\partial y} = -1 \\
\frac{\partial u}{\partial x} &\neq \frac{\partial v}{\partial y}
\end{aligned}
$$

and hence $f(z)$ is not analytic and therefore $f'(z)$ does not exist.

Example 2

Show that u is harmonic in some domain and find a harmonic conjugate v when $u(x, y) = 2x - x^3 + 3xy^2$. Obtain the analytic function $f(z) = u + iv$ as a function of z.

Solution

To show that u is harmonic, i.e u satisfies Laplace's equation,

$$\frac{\partial^2 u}{\partial x^2} + \frac{\partial^2 u}{\partial y^2} = 0$$

Now

$$\frac{\partial u}{\partial x} = 2 - 3x^2 + 3y^2$$

$$\frac{\partial^2 u}{\partial x^2} = -6x$$

$$\frac{\partial u}{\partial y} = 6xy$$

$$\frac{\partial^2 u}{\partial y^2} = 6x$$

So $\quad \dfrac{\partial^2 u}{\partial x^2} + \dfrac{\partial^2 u}{\partial y^2} = 0$

Hence $u(x,y)$ is a harmonic function. Now to find the harmonic conjugate v, we shall use the Cauchy-Riemann equations.

We have

$$\frac{\partial u}{\partial x} = 2 - 3x^2 + 3y^2$$

But by Cauchy-Riemann conditions we have

$$\frac{\partial u}{\partial x} = \frac{\partial v}{\partial y}$$

$$\frac{\partial u}{\partial y} = -\frac{\partial v}{\partial x}$$

Therefore

$$\frac{\partial v}{\partial y} = 2 - 3x^2 + 3y^2$$

Integrating partially with respect to y,

$$v(x,y) = 2y - 3x^2 y + y^3 + \phi(x)$$

and differentiating with respect to x partially we have

$$\frac{\partial v}{\partial x} = -6xy + \phi'(x) = -\frac{\partial u}{\partial y}$$

$$= -6xy$$

So $\quad \phi'(x) = 6xy - 6xy = 0$

Integrating we obtain

$$\phi(x) = c$$

So $\quad v(x,y) = 2y - 3x^2 y + y^3 + c$

is the harmonic conjugate of $u(x,y)$ where c is an arbitrary real constant. The corresponding analytic function is

$$f(z) = u + iv = 2x - x^3 + 3xy^2 + i(2y - 3x^2 y + y^3 + c)$$

It is easily verified that

$$f(z) = 2z - z^3 + ic$$

This form is suggested by noting that when $y = 0$, the analytic function becomes $f(x) = 2x - x^3 + ic$.

Example 3

Sketch the families of level curves of the component functions u and v when $f(z) = z^2$.

Solution

Given that

$$
\begin{aligned}
f(z) &= z^2 \\
&= (x + iy)^2 \\
&= x^2 - y^2 + 2ixy
\end{aligned}
$$

Hence we have

$$
\begin{aligned}
u(x, y) &= x^2 - y^2 \\
v(x, y) &= 2xy
\end{aligned}
$$

It can be easily shown that the Cauchy-Riemann conditions are satisfied by this function. Because

$$
\frac{\partial u}{\partial x} = 2x \qquad\qquad \frac{\partial u}{\partial y} = -2y
$$

$$
\frac{\partial v}{\partial x} = 2y \qquad\qquad \frac{\partial v}{\partial y} = 2x
$$

Therefore

$$
\frac{\partial u}{\partial x} = \frac{\partial v}{\partial y}; \qquad \frac{\partial u}{\partial y} = -\frac{\partial v}{\partial x}
$$

and hence $f(z)$ is analytic everywhere and $f'(z)$ exists which is $2z$.

Now the level curves are

$$
\begin{aligned}
u(x, y) &= x^2 - y^2 = c_1 \\
v(x, y) &= 2xy = c_2
\end{aligned}
$$

where c_1 and c_2 are constants. The slope of the first set of level curves is given by

$$
2x - 2y \left(\frac{dy}{dx} \right)_1 = 0
$$

$$
\left(\frac{dy}{dx} \right)_1 = \frac{x}{y}
$$

Also the slope of the second set of level curves is obtained as

$$
2y + 2x \left(\frac{dy}{dx} \right)_2 = 0
$$

$$
\left(\frac{dy}{dx} \right)_2 = -\frac{y}{x}
$$

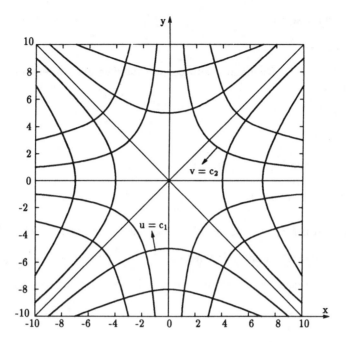

Fig. 1.11: Orthogonal curves of $u = c_1$ and $v = c_2$.

Therefore the product of these two level curves is

$$\left(\frac{dy}{dx}\right)_1 \left(\frac{dy}{dx}\right)_2 = \left(\frac{x}{y}\right)\left(-\frac{y}{x}\right) = -1$$

Since the product is -1, hence the level curves are **orthogonal**. Fig. 1.11 shows the sketch of these two level curves.

Remark

Harmonic functions play a very important role in the solution of the potential problem such as velocity potential in fluid mechanics, electrostatic potential in electric field and the temperatures in thin conducting plates. The use of the theory of complex variables in obtaining such harmonic functions is developed in considerable detail later in Chapter 7.

Exercises

Using Cauchy-Riemann conditions, show that $f'(z)$ does not exist at any point if

1. $f(z) = z^*$

2. $f(z) = z^2 - z^*$

3. $f(z) = 2y - 3ix^2y$

4. $f(z) = e^z + e^{z^*}$

Show that the Cauchy-Riemann conditions are satisfied everywhere for the following $f(z)$ functions:

5. $f(z) = z^2$

6. $f(z) = e^{-x-iy}$

7. $f(z) = \cos x \cosh y - i \sin x \sinh y$

8. At what points does $f(z) = \dfrac{z+1}{(z-1)(z^2+1)}$ fail to be analytic?

9. Verify that if $f(z) = z^3$, then $Re(z^3)$ and $Im(z^3)$ both satisfy Laplace's equation.

10. Does there exist an analytic function $f(z) = u + iv$ for which $u = y^2 + x$?

11. Show that u is a harmonic function and find a harmonic conjugate function v when

   (a) $u(x,y) = e^x \cos y$
   (b) $u(x,y) = 2x(3-y)$
   (c) $u(x,y) = 2y(1-x)$
   (d) $u(x,y) = 2y - y^3 + 3yx^2$
   (e) $u(x,y) = -\cosh x \cos y$

   Obtain the corresponding analytic function $f(z)$ in terms of z for each case.

12. If the analytic function $f(z) = u(x,y) + iv(x,y)$ is expressed in terms of the polar coordinates r and θ, show that the Cauchy-Riemann conditions can be written as

$$\frac{\partial u}{\partial r} = \frac{1}{r}\frac{\partial v}{\partial \theta} \quad \text{and} \quad \frac{\partial v}{\partial r} = -\frac{1}{r}\frac{\partial u}{\partial \theta}$$

   [Hint: Follow the steps demonstrated in Theorem 1.1.]

13. Using the results of Exercise 12, show that when an analytic function is expressed in terms of polar coordinates, both its real and imaginary parts satisfy Laplace's equation in polar coordinates,

$$\frac{\partial^2 \phi}{\partial r^2} + \frac{1}{r}\frac{\partial \phi}{\partial r} + \frac{1}{r^2}\frac{\partial^2 \phi}{\partial \theta^2} = 0$$

14. Show that $u(r,\theta) = \ln r$ is harmonic in the domain $r > 0$, $0 < \theta < 2\pi$. Then find a harmonic conjugate $v(r,\theta)$.

15. Sketch the families of level curves of the component functions u and v when $f(z) = z + 1/z$.

16. Sketch the families of level curves of the component functions u and v when $f(z) = (z+1)/(z-1)$.

1.5 Elementary functions of z: exponential, trigonometric, hyperbolic and logarithmic functions

We shall now consider various elementary functions studied in the calculus of real variables and in this section the corresponding functions of complex variables will be defined. The fundamental of all these complex functions is the exponential function e^z. This function will then be defined in accordance with the definition of the real exponential function e^x.

The following three conditions are posed, which must be satisfied by e^z:

1. e^z shall be single-valued and analytic.

2. $(d/dz)e^z = e^z$.

3. $e^z = e^x$ provided $Im(z) = 0$.

Let us consider that

$$f(z) = e^z = u + iv \tag{1.50}$$

According to condition (1) we have that $f(z)$ is an analytic function of z and hence

$$f'(z) = \frac{\partial u}{\partial x} + i\frac{\partial v}{\partial x} \tag{1.51}$$

But according to condition (2), we have

$$\frac{\partial u}{\partial x} + i\frac{\partial v}{\partial x} = u + iv \tag{1.52}$$

such that

$$\frac{\partial u}{\partial x} = u \tag{1.53}$$

$$\frac{\partial v}{\partial x} = v$$

Now integrating the first part of equation (1.53) partially with respect to x, we get $\ln u = x + \ln \lambda(y)$ which simplifies to $u = e^x \lambda(y)$.

In a similar manner we can integrate the second part of equation (1.53) yielding $\ln v = x + \ln \mu(y)$ so that $v = e^x \mu(y)$ where $\lambda(y)$ and $\mu(y)$ are any arbitrary functions of y.

In view of the analyticity of $f(z)$, Cauchy-Riemann conditions must be satisfied. Therefore,

$$\frac{\partial u}{\partial x} = \frac{\partial v}{\partial y}$$

and

$$\frac{\partial u}{\partial y} = -\frac{\partial v}{\partial x}$$

yield

$$\lambda(y) = \mu'(y)$$
$$\lambda'(y) = -\mu(y)$$

which can be combined to yield one ordinary differential equation

$$\lambda''(y) + \lambda(y) = 0 \qquad (1.54)$$

Solving this equation, we obtain

$$\lambda(y) = A\cos y + B\sin y$$

and hence

$$\mu(y) = A\sin y - B\cos y$$

Thus, we get

$$\begin{aligned} f(z) &= e^z = u + iv \\ &= e^x[(A\cos y + B\sin y) + i(A\sin y - B\cos y)] \end{aligned} \qquad (1.55)$$

But according to condition (3), when $y = 0$, we must have

$$e^x = e^x(A - iB)$$

and this will be true if $A = 1$ and $B = 0$, so that $\lambda(y) = \cos y$ and $\mu(y) = \sin y$ and therefore

$$\begin{aligned} f(z) &= e^x(\cos y + i\sin y) \\ &= e^{x+iy} \\ &= e^z \end{aligned} \qquad (1.56)$$

Thus this is the definition of e^z which meets all the three conditions. It is to be noted that (1.56) is the polar representation of e^z. Here

$$|e^z| = e^x$$

and

$$\arg(e^z) = y$$

Also when $x = 0$, and $y = \theta$, (1.56) can be written as

$$e^{i\theta} = \cos\theta + i\sin\theta \qquad (1.57)$$

and thus

$$re^{i\theta} = r(\cos\theta + i\,\sin\theta) \tag{1.58}$$

Similarly, from (1.57), we have

$$\begin{aligned} e^{-i\theta} &= \cos(-\theta) + i\,\sin(-\theta) \\ &= \cos\theta - i\,\sin\theta \end{aligned} \tag{1.59}$$

Thus using (1.57) and (1.59), we have,

$$\cos\theta = \frac{e^{i\theta} + e^{-i\theta}}{2} \tag{1.60}$$

$$\sin\theta = \frac{e^{i\theta} - e^{-i\theta}}{2i}$$

These two formulae are called Euler's formulae.

Extending this idea to the complex domain, we have,

$$\cos z = \frac{e^{iz} + e^{-iz}}{2} \tag{1.61}$$

$$\sin z = \frac{e^{iz} - e^{-iz}}{2i}$$

From these definitions, it is easy to establish the following familiar formulas:

$$\cos^2 z + \sin^2 z = 1$$
$$\cos(z_1 \pm z_2) = \cos z_1 \cos z_2 \mp \sin z_1 \sin z_2$$
$$\sin(z_1 \pm z_2) = \sin z_1 \cos z_2 \pm \sin z_2 \cos z_1$$
$$\frac{d}{dz}(\cos z) = -\sin z$$
$$\frac{d}{dz}(\sin z) = \cos z$$

Also

$$\tan z = \frac{\sin z}{\cos z}$$

$$\cot z = \frac{\cos z}{\sin z}$$

$$\sec z = \frac{1}{\cos z}$$

$$\csc z = \frac{1}{\sin z}$$

Note that $\tan z$, $\sec z$ are analytic in any domain where $\cos z \neq 0$, and $\cot z$ and $\csc z$ are analytic in any domain where $\sin z \neq 0$. By differentiating, we obtain

$$\frac{d}{dz}(\tan z) = \sec^2 z$$

$$\frac{d}{dz}(\cot z) = -\csc^2 z \qquad\qquad (1.62)$$

$$\frac{d}{dz}(\sec z) = \sec z \tan z$$

$$\frac{d}{dz}(\csc z) = -\csc z \cot z$$

From the Euler definition of $\cos z$ it follows that

$$\cos z = \frac{1}{2}[e^{iz} + e^{-iz}]$$

$$= \frac{1}{2}[e^{i(x+iy)} + e^{-i(x+iy)}]$$

$$= \frac{1}{2}[e^{-y}e^{ix} + e^{y}e^{-ix}]$$

$$= \frac{1}{2}[e^{-y}(\cos x + i \sin x) + e^{y}(\cos x - i \sin x)]$$

$$= \left[\cos x \left(\frac{e^{-y} + e^{y}}{2}\right) + i \sin x \left(\frac{e^{-y} - e^{y}}{2}\right)\right]$$

This can be written in the following form by using the definition of the hyperbolic functions in real variables where $\cosh y = \frac{1}{2}(e^{y} + e^{-y})$ and $\sinh y = \frac{1}{2}(e^{y} - e^{-y})$,

$$\cos z = \cos x \cosh y - i \sin x \sinh y$$

Also from the trigonometric identity we have

$$\cos z = \cos(x + iy)$$
$$= \cos x \cos iy - \sin x \sin iy \qquad\qquad (1.63)$$

Similarly, it can be shown by using the Euler definition of $\sin z$ that

$$\sin z = \sin x \cosh y + i \cos x \sinh y$$

Also from the trigonometric identity we have

$$\sin z = \sin(x + iy)$$
$$= \sin x \cos iy + \cos x \sin iy \qquad\qquad (1.64)$$

Thus it can be easily verified by considering the right-hand side of these identities that

$$\cos iy = \cosh y \qquad\qquad (1.65)$$
$$\sin iy = i \sinh y$$

The hyperbolic functions of z we simply define by extending the familiar definition into complex domain;

$$\cosh z = \frac{e^z + e^{-z}}{2} \tag{1.66}$$

$$\sinh z = \frac{e^z - e^{-z}}{2}$$

The real and imaginary parts of $\cosh z$ and $\sinh z$ can be obtained as

$$\cosh z = \cosh x \cos y + i \sinh x \sin y \tag{1.67}$$

$$\sinh z = \sinh x \cos y + i \sin y \cosh x$$

If $x = 0$, then we find

$$\cosh iy = \cos y \tag{1.68}$$

$$\sinh iy = i \sin y$$

Other hyperbolic functions are defined

$$\tanh z = \frac{\sinh z}{\cosh z}$$

$$\coth z = \frac{\cosh z}{\sinh z}$$

$$\sec h\, z = \frac{1}{\cosh z}$$

$$\csc h\, z = \frac{1}{\sinh z}$$

The remaining hyperbolic functions are defined in terms of $\cosh z$ and $\sinh z$ via the usual identities.

The logarithm of z is defined as the function $w = \ln$ which satisfies the equation

$$z = e^w \tag{1.69}$$

If we let $w = u + iv$ and $z = re^{i\theta}$, then $re^{i\theta} = e^{u+iv} = e^u e^{iv}$ where $r = e^u$ and $\theta = v$. Thus

$$\begin{aligned} w = u + iv &= \ln(re^{i\theta}) \\ &= \ln r + i\theta \\ &= \ln|z| + i \arg z \end{aligned} \tag{1.70}$$

If Θ denotes the principal value of $\arg z$ $(-\pi < \Theta < \pi)$, we can write

$$\theta = \Theta + 2n\pi, \qquad (n = 0, \pm 1, \pm 2, \cdots)$$

This is always true because $e^{i\theta} = e^{i(\Theta + 2n\pi)}$. Thus (1.70) becomes

$$w = \ln|z| + i(\Theta + 2n\pi) \qquad n = 0, \pm 1, \pm 2, \cdots \tag{1.71}$$

which shows that the logarithmic function is infinitely many valued. For any particular value of n a unique branch of the function is determined, and the logarithmic becomes effectively single-valued. When $n = 0$, the resulting formula of the logarithmic function is called the *principal value*. For every n, the corresponding branch of ln z is obviously discontinuous at $z = 0$. Similarly, ln z is discontinuous on the negative real axis. Thus each branch of ln z is continuous and analytic at all points except the points on negative real axis.

From the definition,

$$\ln z = \ln |z| + i \arg z$$
$$= \frac{1}{2}\ln (x^2 + y^2) + i \tan^{-1}\left(\frac{y}{x}\right)$$

it is easy to verify that the Cauchy-Riemann equations are satisfied everywhere except at the origin. Thus,

$$u = \frac{1}{2}\ln (x^2 + y^2)$$
$$v = \tan^{-1}\left(\frac{y}{x}\right)$$

We know that

$$\frac{d}{dz}(\ln z) = \frac{\partial u}{\partial x} + i\frac{\partial v}{\partial x}$$
$$= \frac{x}{x^2 + y^2} - i\frac{y}{x^2 + y^2}$$
$$= \frac{x - iy}{x^2 + y^2}$$
$$= \frac{\bar{z}}{z\bar{z}} = \frac{1}{z}$$

The familiar laws of logarithms of real quantities can be extended to complex quantities:

$$\ln (z_1 z_2) = \ln z_1 + \ln z_2$$
$$\ln \left(\frac{z_1}{z_2}\right) = \ln z_1 - \ln z_2$$
$$\ln z^n = n \ln z$$

Other identities can be obtained referring to the standard textbooks listed in the bibliography.

Example 1

Find all the values of z such that

(a) $e^z = -2$

(b) $e^z = 1 + i\sqrt{3}$

Solution

(a) Given that $e^z = -2$ which can be written as

$$e^{(z \pm 2\pi i)} = -2$$

This can subsequently be written as

$$e^{(z \pm 2n\pi i)} = -2 \qquad n = 0, 1, 2, \cdots$$

Taking logarithm of both sides

$$
\begin{aligned}
\ln \{e^{(z \pm 2n\pi i)}\} &= \ln(-2) \\
z \pm 2n\pi i &= \ln(-1) + \ln 2 \\
&= \ln \{e^{\pm \pi i}\} + \ln 2 \\
&= \pm \pi i + \ln 2 \\
z = \ln 2 &\pm (2n+1)\pi i \qquad n = 0, 1, 2, \ldots
\end{aligned}
$$

(b) Given that

$$e^z = 1 + i\sqrt{3}$$

which can be written as $e^{(z \pm 2n\pi i)} = 1 + i\sqrt{3}$.

Taking logarithm of both sides we obtain

$$
\begin{aligned}
z \pm 2n\pi i &= \ln(1 + i\sqrt{3}) \\
&= \ln 2 + i\,tan^{-1}(\sqrt{3}) \\
&= \ln 2 + i\frac{\pi}{3}
\end{aligned}
$$

Therefore

$$z = \ln 2 + i\pi \left(\frac{1}{3} \pm 2n\right)$$

Example 2

Find all the roots of the equation $\cos z = 2$.

Solution

Given that

$$
\begin{aligned}
\cos z &= 2 \\
\cos(x + iy) &= 2 \\
\cos x \cosh y - i \sinh y \sin x &= 2
\end{aligned}
$$

Therefore, equating real and imaginary parts:

$$\cos x \cosh y = 2$$
$$\sin x \sinh y = 0$$

The second equation gives that $x = n\pi$, $\sinh y \neq 0$. But $\cos n\pi \cosh y = 2$ is not true. So x must be equal to $2n\pi$ where n takes the natural values. Then $\cosh y = 2$ and consequently $y = \cosh^{-1} 2$. Hence we have, $z = x + iy = 2n\pi + i \cosh^{-1} 2$. Alternately we can write the value of $\cosh^{-1} 2$ as follows:

$$\cosh y = 2$$
$$\frac{e^y + e^{-y}}{2} = 2$$
$$e^y + e^{-y} = 4$$
$$e^{2y} - 4e^y + 1 = 0$$

Solving we obtain

$$e^y = \frac{4 \pm \sqrt{16 - 4}}{2} = 2 \pm \sqrt{3}$$

and hence $y = \ln\left(2 \pm \sqrt{3}\right)$ and therefore $z = 2n\pi + i \ln\left(2 \pm \sqrt{3}\right)$ which is the same as $z = 2n\pi + i \cosh^{-1} 2$.

Example 3

Find all the roots of equations

(a) $\cosh z = \dfrac{1}{2}$

(b) $\sinh z = 2$

Solution

(a) Given that

$$\cosh z = \frac{1}{2}$$

Then we have

$$\cosh(x + iy) = \frac{1}{2}$$
$$\cosh x \cosh(iy) + \sinh x \sinh(iy) = \frac{1}{2}$$

or $\cosh x \cos y + i \sinh x \sin y = \dfrac{1}{2}$

Equating real and imaginary parts:

$$\cosh x \cos y = \frac{1}{2}$$
$$\sinh x \sin y = 0$$

From the second relation,

$$\sinh x \neq 0, \qquad \sin y = 0$$

If $y = 2n\pi$ then $\cos(2n\pi) = 1$ and hence

$$\cosh x = \frac{1}{2}$$
$$\cosh x = \frac{1}{2} = \cos\left(\pm\frac{\pi}{3}\right)$$
$$\cos(ix) = \cos\left(\pm\frac{\pi}{3}\right)$$

Thus we obtain

$$ix = \pm\frac{\pi}{3}$$
$$x = \pm i\frac{\pi}{3}$$
$$z = x + iy$$
$$= \left(2n \pm \frac{1}{3}\right)\pi i \qquad n = 0, \pm1, \pm2, \pm3, \ldots$$

(b) Given that

$$\sinh z = i$$
$$\sinh(x + iy) = i$$
$$\sinh x \cosh(iy) + \cosh x \sinh(iy) = i$$
$$\sinh x \cos y + i \cosh x \sin y = i$$

Equating real and imaginary parts:

$$\cos y \sinh x = 0$$
$$\sin y \cosh x = 1$$

From the first one we have $\cos y = 0$ so $y = \pi/2$. Now if $y = \pi/2$ then $\sin y = 1 = \sin\frac{\pi}{2} = 1$.
Hence from the second equation

$$\cosh x = 1$$
$$\cos(ix) = \cos(2n\pi)$$

from which we obtain $ix = \pm 2n\pi$ so that $x = 2n\pi i$ $\qquad n = 0, \pm1, \pm2, \ldots$. Thus $z = x + iy = (2n + \frac{1}{2})\pi i$.

Example 4

Find all the roots of the equation $e^z = -3$.

Solution

Given that

$$e^z = -3$$

Taking the logarithm of both sides

$$
\begin{aligned}
z &= \ln(-3) \\
&= \ln(-1) + \ln 3 \\
&= \ln(e^{\pi i}) + \ln 3 \\
&= \ln[e^{(2n+1)\pi i}] + \ln 3 \\
&= \ln 3 + (2n+1)\pi i \qquad n = 0, \pm 1, \pm 2, \ldots
\end{aligned}
$$

Exercises

Prove the following trigonometric identities:

1. $\cos^2 z + \sin^2 z = 1$

2. $\cos(z_1 \pm z_2) = \cos z_1 \cos z_2 \mp \sin z_1 \sin z_2$

3. $\sin(z_1 \pm z_2) = \sin z_1 \cos z_2 \pm z_2 \cos z_1$

4. $\dfrac{d}{dz}(\sin z) = \cos z$

5. Show that if $z = r e^{i\theta}$, then

   (a) $\bar{z} = r e^{-i\theta}$

   (b) $\exp(\ln r + i\theta) = z$

   (c) $\exp(-nz) = \dfrac{1}{(\exp(z))^n}, \quad (n = 1, 2, \cdots)$

6. Express each of the following functions in the form $a + ib$ where a and b are real numbers:

   (a) $\cos(1 + i)$

   (b) $\cosh(1 + i)$

   (c) $\sinh(2 - i)$

   (d) $\sin^{-1} 3$

   (e) $\tan i$

(f) $\tanh^{-1} 2$

(g) $\ln(-2 + 3i)$

(h) 4^{-i}

(i) $(1 + i)^{1-i}$

7. Find all the values of z such that

(a) $e^z = -1$

(b) $e^z = 2 + \sqrt{3}i$

(c) $\exp(z - 2) = i$

(d) $\sinh z = 2$

(e) $\cosh z = -1$

8. Examine the behaviour of the following functions:

(a) $f(z) = e^z$ when $x \to -\infty$

(b) $f(z) = e^{2z}$ when $y \to \infty$

9. Prove that the function $f(z) = e^{\bar{z}}$ is not analytic anywhere.

10. Prove that $f(z) = e^{1/z}$ is analytic everywhere except the origin.

11. Find the roots of each of the following equations:

(a) $\sin z = \cosh 2$

(b) $\cos z = 4$

(c) $\sin z = \cosh 2$

(d) $\sin z = \cosh 2$

12. Show that $\tan z = \dfrac{\sin 2x + i \sinh 2y}{\cos 2x + \cosh 2y}$

13. Show that $\tanh z = \dfrac{\sinh 2x + i \sin 2y}{\cosh 2x + \cos 2y}$

14. If $f(z) = \sin^{-1} z$, what are the equations of the level curves $u(x, y) = c_1$ and $v(x, y) = c_2$?

15. If $f(z) = \cosh z$, what are the equations of the level curves $u(x, y) = c_1$ and $v(x, y) = c_2$?

16. Show that

(a) $f(z) = \ln(z - i)$ is analytic everywhere except $x \le 0$, $y = 1$

(b) $f(z) = \dfrac{\ln(z+4)}{z^2+i}$ is analytic everywhere except at the points $\pm(1-i)/\sqrt{2}$

and $\quad x \le -4, \quad y = 0$

17. Find the principal value of

(a) i^i

(b) $[1+i]^{2i}$

(c) $\left[\left(\dfrac{e}{2}\right)(1-i\sqrt{e})\right]^{3\pi i}$

Chapter 2

Complex integration

2.1 Introduction

Line integrals $\int f(z)dz$ in the complex plane are defined as follows. Let $f(z) = u(x,y) + iv(x,y)$ be any continuous function of z and let C be any path from A to B in the complex plane along which the integration of the function $f(z)$ is to be performed (Fig. 2.1).

Divide C into n smaller arcs Δs_k by the points $z_k(k = 1, 2, \ldots, n-1)$ and let Δz_k be the infinitesimal chord determined by Δz_k. Choose an arbitrary point ξ_k in each segment Δz_k on C, where $\xi_k = x_k + iy_k$. Then we write

$$\int_C f(z)dz = \lim_{n \to \infty} \sum_{k=1}^{n} f(\xi_k)\Delta z_k \tag{2.1}$$

This means as $n \to \infty$ in such a way that $\Delta z_k \to 0$ and then this integral is called the line integral of $f(z)$ along C if the limit exists.

When the path of integration is a closed curve, in that case the integral is denoted by $\oint_C f(z)dz$ and is often called the contour integration.

It is an important property to mention that

$$|\sum_{k=1}^{n} f(\xi_k)\Delta z_k| \leq \sum_{k=1}^{n} |f(\xi_k)\Delta z_k|$$

$$= \sum_{k=1}^{n} |f(\xi_k)||\Delta z_k|$$

Thus from the relation (2.1) when $n \to \infty$ we have that

$$\left| \int_C f(z)dz \right| \leq \int_C |f(z)||dz| \tag{2.2}$$

or $\qquad \left| \int_C f(z)dz \right| \leq \int_C |f(z)||dz|$

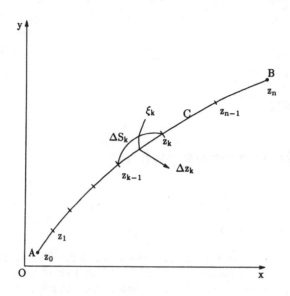

Fig. 2.1: Complex line integral.

If $f(z)$ is a bounded function of z then there exists a constant M such that $|f(z)| \leq M$. Then

$$\left| \int_C f(z)dz \right| \leq \int_C |f(z)||dz|$$

$$\leq \int_C M|dz|$$

$$= ML$$

where $\int_C dz = L =$ length of the arc. Therefore we obtain an important inequality

$$\left| \int_C f(z)dz \right| \leq ML \tag{2.3}$$

where M is any bound for $|f(z)|$ and L is the length of path of integration.

If we take the real and imaginary parts in (2.1), we find

$$\int_C f(z)dz = \lim_{n \to \infty} \sum_{k=1}^{n}(u+iv)(\Delta x + i\Delta y)$$

$$= \lim_{n \to \infty} \left[\sum_{k=1}^{n}(u\Delta x - v\Delta y) + i\sum_{k=1}^{n}(u\Delta y + v\Delta x) \right]$$

that is,

$$
\begin{aligned}
\int_C f(z)dz &= \int_C (u + iv)(dx + idy) \\
&= \int_C (u\,dx - v\,dy) + i \int_C (u\,dy + v\,dx)
\end{aligned}
\tag{2.4}
$$

The complex line integral is thus simply a combination of two real line integrals. Hence, we can apply all the theory of real line integrals. Now from the known properties of real line integrals, we can write

$$
\int_A^B f(z)\,dz = - \int_A^B f(z)\,dz
\tag{2.5}
$$

$$
\int_A^B \lambda f(z)\,dz = \lambda \int_A^B f(z)\,dz
\tag{2.6}
$$

$$
\int_A^B [f(z) \pm g(z)]dz = \int_A^B f(z)\,dz \pm \int_A^B g(z)\,dz
\tag{2.7}
$$

and
$$
\int_A^B f(z)\,dz = \int_A^P f(z)\,dz + \int_P^B f(z)\,dz
\tag{2.8}
$$

where P is a point on the arc AB.

In the following we shall consider some important theorems concerning the complex integration. Before we do that, we first define simply and multiply connected domains which are of paramount importance for establishing these theorems.

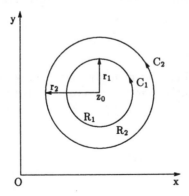

Fig. 2.2: Simply and multiply connected regions.

A simply connected domain R is a domain such that every simple closed contour within it encloses only points of R. As for example, the set of points interior to a simple closed contour is the simply connected domain.

A multiply connected domain is a domain which is not simply connected. As for example, the annular domain between two concentric circles is the multiply connected domain and multiply connected regions.

These ideas are demonstrated in Fig. 2.2 where R_1 and R_2 are two domains defined as

$$R_1: \qquad |z - z_0| \leq r_1$$
$$R_2: \qquad r_1 < |z - z_0| \leq r_2$$

Here C_1 encloses all the points of the domain R_1 and hence R_1 is defined as the simply connected domain. On the other hand, C_2 encloses all the points of R_2 and R_1, therefore domain R_2 is defined as the multiply connected domain.

2.2   The Cauchy-Goursat theorem

Theorem 2.1

If R is a region, either simply or multiply connected, whose boundary is a closed contour C and if $P(x,y)$, $Q(x,y)$, $\partial P/\partial y$ and $\partial Q/\partial x$ are continuous within and on the boundary of R, then

$$\int_C (Pdx + Qdy) = \int\int_R \left(\frac{\partial Q}{\partial x} - \frac{\partial P}{\partial y}\right) dxdy$$

where the integration is taken around C in the positive direction with respect to the interior of R. This theorem is known as Green's lemma and is referred to in many textbooks.

Theorem 2.2

If R is a region, either simply or multiply connected, whose boundary is a closed contour C and if $f(z)$ is analytic and $f'(z)$ is continuous within and on the boundary of R, then

$$\int_C f(z)dz = 0$$

Proof

We know that $f(z)$ can be written as $f(z) = u + iv$ and $dz = dx + idy$.
Thus

$$\int_C f(z)dz = \int_C (u + iv)(dx + idy)$$
$$= \int_C (udx - vdy) + i\int_C (udy + vdx)$$

Now by the assumption that $f'(z)$ exists and is continuous means that partial derivatives u_x, u_y, v_x and v_y exist and are continuous throughout R. Hence, by Green's lemma, we have

$$\int_C f(z)\, dz = \int\int_R \left(-\frac{\partial v}{\partial x} - \frac{\partial u}{\partial y}\right) dx dy + i \int\int_R \left(\frac{\partial u}{\partial x} - \frac{\partial v}{\partial y}\right) dx dy$$

Now applying Cauchy-Riemann conditions

$$\frac{\partial u}{\partial x} = \frac{\partial v}{\partial y}$$

$$\frac{\partial u}{\partial y} = -\frac{\partial v}{\partial x}$$

into the right-hand side of the above relation, we have $\int_C f(z)dz = 0$. Hence this is the required proof.

This theorem is usually attributed to the French mathematicians, Augustin Louis Cauchy (1789-1857) and Edouard Goursat (1858-1936) and is called Cauchy-Goursat Theorem.

Remark

The simple closed contour here can be replaced by an arbitrary closed contour C which is not necessarily simple. In particular, if $f(z)$ is analytic within and on the boundary of the region R between two simple closed curves as shown in Fig. 2.3 (a) and (b), then from the Cauchy-Goursat theorem:

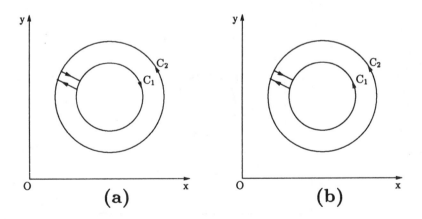

Fig. 2.3: (a) Contours C_1 and C_2 in the opposite directions. (b) Contours C_1 and C_2 in the positive directions.

$$\int_{C_1} f(z)dz \; + \; \int_{C_2} f(z)dz = 0$$

or $$\int_{C_1} f(z)dz \; = \; \int_{C_2} f(z)dz$$

These two results are obtained considering the contour C_1 and C_2 in the opposite directions and in the positive directions, respectively. One of the most striking consequences of the Cauchy-Goursat theorem is **Cauchy's integral formula** which will be established below.

Theorem 2.3

If $f(z)$ is analytic everywhere within and on a simple closed contour C, taken in the positive sense (anticlockwise direction) and if z_0 is any point interior to C, then

$$f(z_0) = \frac{1}{2\pi i} \int_C \frac{f(z)}{z - z_0} dz$$

This formula is known as the **Cauchy's integral formula**.

Proof

Let C_0 be a circle $|z - z_0| = r_0$ about z_0 where the radius r_0 is small enough so that C_0 is interior to C (see Fig. 2.4).

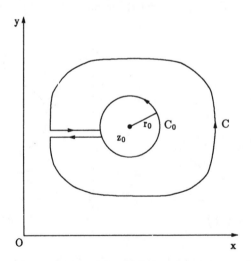

Fig. 2.4: Cauchy's integral formula.

Hence the function $f(z)/(z - z_0)$ is analytic at all points within and on C except at the point z_0. Hence, by **Cauchy-Goursat theorem**, the integral of this function around C is equal to its integral around C_0. Thus

$$
\begin{aligned}
\int_C \frac{f(z)}{z - z_0} dz &= \int_{C_0} \frac{f(z)}{z - z_0} dz \\
&= \int_{C_0} \frac{f(z_0) + [f(z) - f(z_0)]}{z - z_0} dz \\
&= f(z_0) \int_{C_0} \frac{dz}{z - z_0} + \int_{C_0} \frac{f(z) - f(z_0)}{z - z_0} dz \quad (2.9)
\end{aligned}
$$

In the first integral if we put $z - z_0 = r_0 e^{i\theta}$ such that $dz = r_0 i e^{i\theta} d\theta$ we obtain

$$
\begin{aligned}
\int_{C_0} \frac{dz}{z - z_0} &= \int_{\theta=0}^{2\pi} \frac{r_0 i e^{i\theta}}{r_0 e^{i\theta}} d\theta \\
&= 2i\pi
\end{aligned}
$$

Considering the second integral we have

$$
\left| \int_{C_0} \frac{f(z) - f(z_0)}{z - z_0} dz \right| \leq \int_{C_0} \frac{|f(z) - f(z_0)|}{|z - z_0|} |dz| \quad (2.10)
$$

Now on C_0 we have $|z - z_0| = r_0$. Also we have that $f(z)$ is analytic and hence continuous, it follows that for any ε there exists a δ such that

$$
|f(z) - f(z_0)| < \varepsilon \qquad \text{whenever} \qquad |z - z_0| = r_0 < \delta
$$

Now inserting these estimates into (2.10) we obtain,

$$
\begin{aligned}
& \left| \int_{C_0} \frac{f(z)-f(z_0)}{z-z_0} dz \right| \\
& \leq \int_{C_0} \left| \frac{f(z)-f(z_0)}{|z-z_0|} \right| |dz| \\
& < \int_{C_0} \frac{\varepsilon}{r_0} |dz| = \frac{\varepsilon}{r_0} (2\pi r_0) = 2\pi\varepsilon
\end{aligned}
$$

In the limiting process, when $\varepsilon \to 0$, the integral

$$
\left| \int_{C_0} \frac{f(z) - f(z_0)}{z - z_0} dz \right| \to 0
$$

Thus

$$
\int_C \frac{f(z)}{z - z_0} dz = 2\pi i \, f(z_0)
$$

and hence

$$f(z_0) = \frac{1}{2\pi i} \int_C \frac{f(z)}{z - z_0} dz$$

Hence this is the required proof.

The Cauchy integral formula can be extended to the case where $f(z)$ is analytic and possesses nth derivatives. The next theorem gives the extension of Cauchy integral formula.

Theorem 2.4

If $f(z)$ is analytic everywhere within and on a simple closed contour C, and if the derivatives of $f(z)$ of all orders exist and are analytic, and if z_0 is any point interior to C, then

$$f^{(n)}(z_0) = \frac{n!}{2\pi i} \int_C \frac{f(z)}{(z - z_0)^{n+1}} dz$$

Proof

We shall first establish the following result corresponding to $n = 1$, i.e., the first derivative, and then extend it to n. By definition

$$
\begin{aligned}
f'(z_0) &= \lim_{\Delta z_0 \to 0} \frac{f(z_0 + \Delta z_0) - f(z_0)}{\Delta z_0} \\
&= \lim_{\Delta z_0 \to 0} \frac{1}{\Delta z_0} \left[\frac{1}{2\pi i} \int_C \frac{f(z)}{z - (z_0 + \Delta z_0)} dz - \frac{1}{2\pi i} \int_C \frac{f(z)}{z - z_0} dz \right] \\
&= \lim_{\Delta z_0 \to 0} \frac{1}{\Delta z_0} \left[\frac{1}{2\pi i} \int_C f(z) \left\{ \frac{1}{z - (z_0 + \Delta z_0)} - \frac{1}{z - z_0} \right\} dz \right] \\
&= \lim_{\Delta z_0 \to 0} \left[\frac{1}{2\pi i} \int_C \frac{f(z) dz}{(z - z_0 - \Delta z_0)(z - z_0)} \right] \\
&= \frac{1}{2\pi i} \int_C \frac{f(z) dz}{(z - z_0)^2} \tag{2.11}
\end{aligned}
$$

Continuing in this way, we obtain the following formulae:

$$
\begin{aligned}
f''(z_0) &= \frac{2!}{2\pi i} \int_C \frac{f(z)}{(z - z_0)^3} dz \\
f'''(z_0) &= \frac{3!}{2\pi i} \int_C \frac{f(z)}{(z - z_0)^4} dz \qquad \text{etc.}
\end{aligned}
$$

and hence

$$f^{(n)}(z_0) = \frac{n!}{2\pi i} \int_C \frac{f(z) dz}{(z - z_0)^{n+1}}$$

Hence this is the required proof.

Beginning with the formula given above we can now establish Cauchy's inequality in the next theorem. Using this theorem we can prove the converse of Cauchy's theorem, which is known as Morera's theorem.

Theorem 2.5

If $f(z)$ is analytic within and on a circle C of radius r with centre at z_0, then

$$|f^{(n)}(z_0)| \leq \frac{n!\, M}{r^n}$$

where M is the maximum value of $f(z)$ on C.

Proof

From the above theorem, we have

$$
\begin{aligned}
|f^{(n)}(z_0)| &= \left| \frac{n!}{2\pi i} \int_C \frac{f(z)dz}{(z-z_0)^{n+1}} \right| \\
&\leq \frac{n!}{2\pi} \int_C \frac{|f(z)|\,|dz|}{|z-z_0|^{n+1}} \\
&\leq \frac{n!}{2\pi} \frac{M}{r^{n+1}} \int_C |dz| \\
&= \frac{n!}{2\pi} \frac{M}{r^{n+1}} 2\pi r \\
&= \frac{n!\, M}{r^n}
\end{aligned}
$$

Hence this is the required proof.

Summary of the Cauchy integral formulae

We formally define the poles and the residue of the function at the poles in Chapter 4. However, at this point we simply define a pole as a singular point at which the function goes to infinity and we will find later that at each pole there exists a residue.

If $f(z)$ is analytic within and on C, where C is a closed contour shown in Fig. 2.5, then the residues at the simple, double, triple and multiple pole of order $(n+1)$ can, respectively, be obtained from the following simple formulae:

(1) Residue at a simple pole $z = z_0$.

$$
\begin{aligned}
\int_C \frac{f(z)}{z-z_0}dz &= 2\pi i \lim_{z \to z_0} \left\{ (z-z_0)\frac{f(z)}{z-z_0} \right\} \\
&= 2\pi i f(z_0) = 2\pi i R_1
\end{aligned}
$$

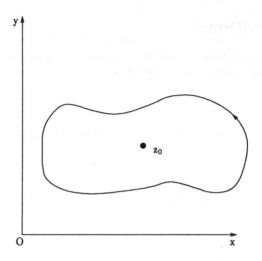

Fig. 2.5: A closed contour.

(2) Residue at a double pole $z = z_0$.

$$\int_C \frac{f(z)}{(z - z_0)^2} dz = \lim_{z \to z_0} \frac{1}{1!} \frac{d}{dz} \left[(z - z_0)^2 \frac{f(z)}{(z - z_0)^2} \right]$$
$$= 2\pi i \, f'(z_0) = 2\pi i \, R_2$$

(3) Residue at a triple pole $z = z_0$.

$$\int_C \frac{f(z)}{(z - z_0)^3} dz = (2\pi i) \lim_{z \to z_0} \frac{1}{2!} \frac{d^2}{dz^2} \left[(z - z_0)^3 \frac{f(z)}{(z - z_0)^3} \right]$$
$$= (2\pi i) \frac{1}{2!} f''(z_0) = \frac{2\pi i}{2!} f''(z_0)$$

(4) Residue at $z = z_0$ a multiple pole of order $(n + 1)$.

$$\int_C \frac{f(z)}{(z - z_0)^{n+1}} dz = (2\pi i) \lim_{z \to z_0} \frac{1}{n!} \frac{d^n}{dz^n} \left[(z - z_0)^{n+1} \frac{f(z)}{(z - z_0)^{n+1}} \right]$$
$$= \frac{2\pi i}{n!} f^{(n)}(z_0)$$

In summary, if $g(z)$ is a function which has a simple pole at $z = z_0$ and this pole is inside the contour C, then

$$\int_C g(z)dz = (2\pi i) \times \lim_{z \to z_0} [(z - z_0)g(z)]$$

For multiple pole of multiplicity $(n + 1)$

$$\int_C g(z)dz = (2\pi i) \times \lim_{z \to z_0} \frac{1}{n!} \frac{d^n}{dz^n} [(z - z_0)^{n+1} g(z)]$$

2.3 Applications

This section contains applications of line integrals in the complex plane. These applications are demonstrated by using the Cauchy integral formulae.

Example 1

Evaluate integral $\int_{C_0} f(z)dz$ where the contour C_0 is the circle $|z - z_0| = r_0$.

Solution

Denoting the integral by I and using the substitution $z = z_0 + r_0 e^{i\theta}$ such that $dz = r_0 i e^{i\theta} d\theta$, we obtain

$$I = \int_0^{2\pi} f(z_0 + r_0 e^{i\theta})(r_0 i e^{i\theta} d\theta)$$

$$= i r_0 \int_0^{2\pi} f(z_0 + r_0 e^{i\theta})e^{i\theta} d\theta$$

This is the required result of the given integral.

Example 2

Evaluate the integral $\int_{C_0} dz/(z - z_0)$ where the contour C_0 is the circle $|z - z_0| = r_0$.

Solution

Denoting the integral by I and using the substitution $z = z_0 + r_0 e^{i\theta}$ such that $dz = r_0 i e^{i\theta} d\theta$, we obtain

$$I = \int_0^{2\pi} \frac{r_0 i e^{i\theta} d\theta}{r_0 e^{i\theta}} = 2\pi i$$

Example 3

Evaluate the following integral:

$$I = \int_{C_0} \frac{dz}{(z - z_0)^3}$$

where the contour C_0 is the circle given by $|z - z_0| = r_0$.

Solution

Here $z = z_0$ is a pole of order three of the given integrand. We substitute $z = z_0 + r_0 e^{i\theta}$ such that $dz = r_0 i e^{i\theta} d\theta$, and we obtain

$$
\begin{aligned}
I &= \int_0^{2\pi} \frac{r_0 i e^{i\theta} d\theta}{r_0^3 e^{3i\theta}} \\
&= \frac{i}{r_0^2} \int_0^{2\pi} e^{-2i\theta} d\theta \\
&= \frac{i}{r_0^2} \left[\frac{e^{-2i\theta}}{-2i} \right]_0^{2\pi} \\
&= -\frac{1}{2r_0^2}[e^{-4\pi i} - 1] = -\frac{1}{2r_0^2}[1 - 1] = 0
\end{aligned}
$$

Example 4

Evaluate the following integral:

$$
\int_C \frac{z^3 + 2z}{(z-2)^3} dz
$$

where C is a closed contour given by the circle $|z| = 3$.

Solution

Using the Cauchy integral formula to find the residue at $z = 2$ the triple pole of the integrand which is inside the given contour and denoting the integral by I we obtain

$$
\begin{aligned}
I &= (2\pi i) \times \lim_{z \to 2} \frac{1}{2} \frac{d^2}{dz^2}(z^3 + 2z) \\
&= (\pi i) \times 12 \\
&= 12\pi i
\end{aligned}
$$

Example 5

Evaluate the following integral:

$$
I = \int_C \frac{2z}{(z-2)^3} dz
$$

where the contour C is given by the circle $|z| = 2$.

Solution

Since $z = 2$ is a pole of order three which is inside the given contour, the residue i.e. the value of the integral is given by (using the Cauchy integral formula)

$$
\begin{aligned}
I &= (2\pi i) \times \lim_{z \to 2} \frac{1}{2} \frac{d^2}{dz^2}(2z) \\
&= 0
\end{aligned}
$$

Example 6

Evaluate the following integral:

$$
I = \int_C \frac{2z}{(z - 2)^2} dz
$$

where the contour C is given by the circle $|z| = 3$.

Solution

Using the Cauchy integral formula for the double pole at $z = 2$ which is inside the contour, we obtain

$$
\begin{aligned}
I &= (2\pi i) \times \lim_{z \to 2} \frac{d}{dz}(2z) \\
&= 4\pi i
\end{aligned}
$$

Example 7

Show that

$$
\oint_C \frac{dz}{z^2(z^2 + 9)} = 0
$$

where the contour C is a unit circle $|z| = 1$.

Proof

Here $z = 0$ is the only pole (a double pole) which is inside the given contour. Hence the value of the integral is obtained by using the Cauchy integral formula.

$$
\begin{aligned}
\oint_C \frac{dz}{z^2(z^2 + 9)} &= (2\pi i) \times \lim_{z \to 0} \frac{d}{dz}\left(\frac{1}{z^2 + 9}\right) \\
&= (2\pi i) \times \lim_{z \to 0} \frac{-2z}{(z^2 + 9)^2} \\
&= 0
\end{aligned}
$$

as asserted.

Note

In summary, if $f(z)$ is an analytic function defined within and on the closed contour C, and $z = z_0$ is a pole of order $(n + 1)$ which is within the given contour then the following results hold good:

Cauchy integral formulae:

$$f^{(n)}(z_0) = \frac{n!}{2\pi i} \oint_C \frac{f(z)}{(z - z_0)^{n+1}} dz \qquad n = 0, 1, 2, 3, \cdots$$

Consider that $f(z) = 1$. Then using the above formula we obtain the following important results:

- Residue at a simple pole $z = z_0$ when $f(z) = 1$

$$\oint_C \frac{dz}{z - z_0} = (2\pi i) \times \lim_{z \to z_0} (1) = 2\pi i$$

- Residue at a double pole $z = z_0$ when $f(z) = 1$

$$\oint_C \frac{dz}{(z - z_0)^2} = (2\pi i) \times \lim_{z \to z_0} \frac{d}{dz}(1) = 0$$

- Residue at $z = z_0$ a multiple pole of order $(n + 1)$ when $f(z) = 1$

$$\oint_C \frac{dz}{(z - z_0)^{n+1}} = \frac{2\pi i}{n!} \lim_{z \to z_0} \frac{d^n}{dz^n}(1) = 0$$

Cauchy-Goursat theorem
If $f(z)$ is analytic within and on C, then

$$\oint_C f(z)dz = 0$$

Example 8

Evaluate the following integrals:

    (a) $\int_C z^2 dz$, where C is the line segment OB from $z = 0$ to $z = 2 + i$.

    (b) $\int_C z^2 dz$, where C consists of two line segments, one from $z = 0$ to $z = 2$ and the other from $z = 2$ to $z = 2 + i$.

    (c) $\int_C z^2 dz$, where C consists of two line segments, one from $z = 0$ to $z = i$, and the other from $z = i$ to $z = 2 + i$.

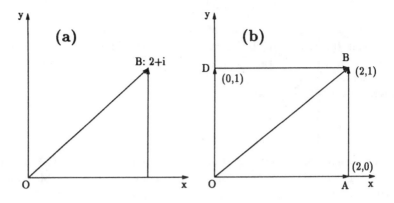

Fig. 2.6: A contour.

Solution

The contour is shown in Fig. 2.6.

(a) In this case, the integration is performed along the contour C which is the line segment $\overrightarrow{OB}$. The equation of the line is $x = 2y$ and therefore, $dx = 2dy$. Also $dz = dx + idy$. Therefore the integral is evaluated as follows:

$$
\begin{aligned}
\int_C z^2 dz &= \int_{\overrightarrow{OB}} (x + iy)^2 (dx + idy) \\
&= \int_{\overrightarrow{OB}} (x^2 - y^2 + 2i\,xy)(dx + i\,dy) \\
&= \int_0^1 (2 + i)(4y^2 - y^2 + 4i\,y^2)\,dy \\
&= (2 + i)(3 + 4i) \int_0^1 y^2 dy \\
&= \frac{1}{3}(2 + i)(3 + 4i) = \frac{2}{3} + \frac{11}{3}i
\end{aligned}
$$

(b) The integration in this case is performed along the contour C which is composed of two line segments $\overrightarrow{OA}$ and $\overrightarrow{AB}$. It is to be noted that along $\overrightarrow{OA}$, x is variable while $y = 0$ such that $dy = 0$. However, along $\overrightarrow{AB}$, y is variable and $x = 2$ such that $dx = 0$. Thus using this observation, the value of the integral is obtained.

$$
\begin{aligned}
\int_C z^2 dz &= \int_{\overrightarrow{OA}} z^2 dz + \int_{\overrightarrow{AB}} z^2 dz \\
&= \int_{\overrightarrow{OA}} (x^2 - y^2 + 2ixy)(dx + idy) + \int_{\overrightarrow{AB}} (x^2 - y^2 + 2ixy)(dx + idy)
\end{aligned}
$$

$$= \int_0^2 x^2 dx + \int_0^1 (4 - y^2 + 4iy)(idy)$$

$$= \frac{8}{3} + i \int_0^1 (4 - y^2) dy - 4 \int_0^1 y dy$$

$$= \frac{8}{3} + \frac{11}{3}i - 2$$

$$= \frac{2}{3} + \frac{11}{3}i$$

(c) The integration in this case is performed along the contour C which is made of two line segments $\overrightarrow{OD}$ and $\overrightarrow{DB}$. We see that along $\overrightarrow{OD}$, y is variable while $x = 0$ such that $dx = 0$. But along $\overrightarrow{DB}$, x is variable and $y = 1$ such that $dy = 0$. Hence performing this integration we have

$$\int_C z^2 dz = \int_C (x + iy)^2 (dx + idy)$$

$$= \int_C (x^2 - y^2 + 2ixy)(dx + idy)$$

$$= \int_{\overrightarrow{OD}} (x^2 - y^2 + 2ixy)(dx + idy) + \int_{\overrightarrow{DB}} (x^2 - y^2 + 2ixy)(dx + idy)$$

$$= \int_0^1 -y^2 idy + \int_0^2 (x^2 - 1 + 2ix) dx$$

$$= -i \left| \frac{y^3}{3} \right|_0^1 + \left[\frac{x^3}{3} - x + ix^2 \right]_0^2$$

$$= -\frac{i}{3} + \left[\frac{8}{3} - 2 + 4i \right]$$

$$= \frac{2}{3} + \frac{11}{3}i$$

It is worth mentioning here that the value of the integral in (a) is equal to the value of the integral in (b), and subsequently the value of the integral along the closed contour $O\overrightarrow{ABO}$ is equal to zero. The reason behind this is that the function $f(z)$ is analytic within and on the closed contour $O\overrightarrow{ABO}$. To see this point we have that

$$u = x^2 - y^2$$
$$v = 2xy$$
$$\frac{\partial u}{\partial x} = 2x = \frac{\partial v}{\partial y}$$
$$\frac{\partial u}{\partial y} = -2y = -\frac{\partial v}{\partial x}$$

Thus the Cauchy-Riemann conditions are satisfied and hence $\int_C f(z) dz = 0$ according to Cauchy-Goursat theorem.

Example 9

Evaluate the following integrals:

(a) $\int_C (y - x - 3x^2 i)dz$, where C is the line segment from $z = 0$ to $z = 1 + i$.

(b) $\int_C (y - x - 3x^2 i)dz$, where C consists of two line segments, one from $z = 0$ to $z = i$, and the other from $z = i$ to $z = 1 + i$.

(c) $\int_C (y - x - 3x^2 i)dz$, where C consists of two line segments, one from $z = 0$ to $z = 1$, and the other from $z = 1$ to $z = 1 + i$.

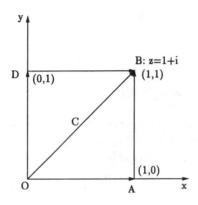

Fig. 2.7: A contour.

Solution

The contour is shown in Fig. 2.7.

(a) In this case the integration is performed along the contour C which is given by the line segment $\overrightarrow{OB}$. From the Fig. 2.7 it is observed that along $\overrightarrow{OB}$, $y = x$ such that $dy = dx$ and consequently, $dz = (1 + i)dx$. Hence the value of the integral is

$$
\begin{aligned}
\int_C (y - x - i3\,x^2)dz &= \int_{\overrightarrow{OB}} (y - x - i\,3x^2)(dx + i\,dy) \\
&= \int_0^1 (-i\,3x^2)(1 + i)dx \\
&= -3i(1 + i) \int_0^1 x^2 dx \\
&= -3(i - 1) \left[\frac{x^3}{3} \right]_0^1 = 1 - i
\end{aligned}
$$

(b) The integration in this case is performed along the contour C which is composed of the line segments $\overrightarrow{OD}$ and $\overrightarrow{DB}$. Now along $\overrightarrow{OD}$, y is variable and $x = 0$ such that

$dx = 0$. Thus the integration yields

$$\int_C (y - x - i\,3x^2)dz = \int_{\overrightarrow{OD}} (y - x - i3x^2)(dx + idy)$$

$$+ \int_{\overrightarrow{DB}} (y - x - i\,3x^2)(dx + idy)$$

$$= \int_0^1 y(idy) + \int_0^1 (1 - x - i\,3x^2)dx$$

$$= i\left(\frac{y^2}{2}\right)_0^1 + \left[x - \frac{x^2}{2} - i\,x^3\right]_0^1$$

$$= \frac{i}{2} + \left(1 - \frac{1}{2} - i\right)$$

$$= \frac{1}{2}(1 - i)$$

(c) In this case the integration is performed along the contour C which is composed of the line segments $\overrightarrow{OA}$ and $\overrightarrow{AB}$. In this case along $\overrightarrow{OA}$, x is variable while $y = 0$ such that $dy = 0$. Also along $\overrightarrow{AB}$, y is variable but $x = 1$ so $dx = 0$.

$$\int_C (y - x - i\,3x^2)(dx + i\,dy) = \int_{\overrightarrow{OA}} (y - x - i3x^2)(dx + idy)$$

$$+ \int_{\overrightarrow{AB}} (y - x - 3ix^2)(dx + idy)$$

$$= \int_0^1 (-x - 3ix^2)dx + \int_0^1 (y - 1 - 3i)(idy)$$

$$= -\frac{1}{2} - i + i\left(\frac{1}{2} - 1 - 3i\right)$$

$$= \frac{5}{2} - \frac{3}{2}i$$

It is to be noted here that the value of the integral in (a) is not equal to that of the integral in (b) and subsequently the value of the integral along the closed contour $O\overrightarrow{ABO}$ is not zero. The reason behind this is that $f(z)$ is not analytic within and on the closed contour $O\overrightarrow{ABO}$. It is evident from the following analysis. Here we have

$$u = y - x, \qquad v = -3x^2$$
$$\frac{\partial u}{\partial x} = -1, \qquad \frac{\partial v}{\partial y} = 0$$
$$\frac{\partial u}{\partial y} = 1, \qquad \frac{\partial v}{\partial x} = -6x$$

If $f(z)$ is analytic, then the Cauchy-Riemann conditions must be satisfied i.e., $(\partial u/\partial x) = (\partial v/\partial y)$ and $(\partial u/\partial y) = -(\partial v/\partial x)$. These two conditions are not satisfied in this case. Hence the integral is not zero within the closed contour.

Example 10

Evaluate the integral $\int_C (z+2)/z \, dz$ along the contour C as shown in Fig. 2.8.

(a) along the semicircle $z = 2e^{i\theta}$ $(0 \le \theta \le \pi)$ $\quad |z| = 2$

(b) along the semicircle $z = 2e^{i\theta}$ $(\pi \le \theta \le 2\pi)$ $\quad |z| = 2$

(c) along the circle $z = 2e^{i\theta}$ $(-\pi \le \theta \le \pi)$

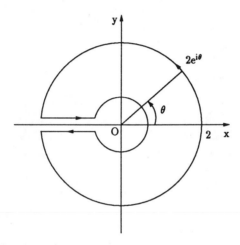

Fig. 2.8: A closed contour C.

Solution

(a) Consider the contour as shown in Fig. 2.8. In this case, $z = 2e^{i\theta}$ such that $dz = 2ie^{i\theta} d\theta$. Hence the value of the integral is

$$
\begin{aligned}
\int_C \left(\frac{z+2}{z} \right) dz &= \int_0^\pi \left(1 + \frac{2}{2e^{i\theta}} \right) (2ie^{i\theta} d\theta) \\
&= 2i \int_0^\pi (1 + e^{-i\theta}) e^{i\theta} d\theta \\
&= 2i \int_0^\pi (e^{i\theta} + 1) d\theta \\
&= 2i \left[\frac{e^{i\theta}}{i} + \theta \right]_0^\pi \\
&= 2i \left[\frac{e^{i\pi}}{i} + \pi - \frac{1}{i} \right] = 2[-2 + \pi i] = -4 + 2\pi i
\end{aligned}
$$

(b) In this case we have

$$\int_C \frac{z+2}{z}dz = \int_\pi^{2\pi} (1 + e^{-i\theta})2ie^{i\theta}d\theta$$

$$= 2i \int_\pi^{2\pi} (1 + e^{i\theta})d\theta$$

$$= 2i \left[\theta + \frac{e^{i\theta}}{i}\right]_\pi^{2\pi}$$

$$= 2i \left[(2\pi - \pi) + \frac{1}{i}(e^{2\pi i} - e^{\pi i})\right]$$

$$= 2i \left[\pi + \frac{1}{i}(1 + 1)\right]$$

$$= 2\pi i + 4$$

(c) In this case we have

$$\int_C \frac{z+2}{z}dz = 2i \int_{-\pi}^{\pi} (1 + e^{i\theta})d\theta$$

$$= 2i \left[\theta + \frac{e^{i\theta}}{i}\right]_{-\pi}^{\pi}$$

$$= 2i \left[2\pi + \frac{1}{i}(e^{i\pi} - e^{-i\pi})\right]$$

$$= 4\pi i + 0$$

$$= 4\pi i$$

Example 11

Prove that

$$\left|\int_C \frac{\ln z}{z^2}dz\right| < 2\pi \frac{\pi + \ln R}{R}$$

where C is the circle $|z| = R > 1$, taken in the anticlockwise direction and hence show that as $R \to \infty$ the value of the integral $\to 0$.

Solution

In the given integral $\int_C (\ln z/z^2)dz$, substitute $z = Re^{i\theta}$ so that $dz = Rie^{i\theta}d\theta$ and $\ln z = \ln R + i\theta$. Then we have

$$\left| \int_C \frac{\ln z}{z^2} dz \right| \leq \int_C \frac{|\ln z|}{|z^2|} |dz|$$

$$= \int_C \frac{|\ln R + i\theta|}{R^2} |dz|$$

$$\leq \int_0^{2\pi} \frac{\ln R + \theta}{R^2} R d\theta$$

$$= \left[\frac{\ln R}{R} \theta + \frac{\theta^2}{2R} \right]_0^{2\pi}$$

$$= \frac{\ln R}{R} 2\pi + \frac{4\pi^2}{2R}$$

$$= 2\pi \left(\frac{\ln R + \pi}{R} \right)$$

as asserted.

Now if $R \to \infty$ then $\lim_{R\to\infty} ((\ln R + \pi)/R) \to 0$. Consequently $\left| \int_C (\ln z/z^2) dz \right| \to 0$ as $R \to \infty$.

Example 12

Find the value of the integral of $g(z)$ around the simple closed contour $|z - i| = 2$

(a) $g(z) = 1/(z^2 + 4)$

(b) $g(z) = 1/(z^2 + 4)^2$ (Use Cauchy integral formula.)

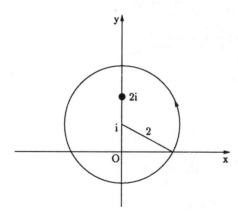

Fig. 2.9: A closed contour.

Solution

(a) The contour C is shown by Fig. 2.9. Here

$$g(z) = \frac{1}{z^2 + 4}$$

$f(z)$ is analytic everywhere except at $z = \pm 2i$.

Thus $z = 2i$ is the point inside the given closed contour C where the function is not analytic. Thus by the Cauchy integral formula

$$
\begin{aligned}
\int_C g(z)dz &= \int_C \frac{dz}{z^2 + 4} \\
&= \int_C \frac{dz}{(z + 2i)(z - 2i)} \\
&= 2\pi i \left[\frac{1}{z + 2i} \right]_{z=2i} = \frac{\pi}{2}
\end{aligned}
$$

(b) By the Cauchy integral formula we have

$$
\begin{aligned}
\int_C g(z)dz &= \int_C \frac{dz}{(z^2 + 4)^2} \\
&= \int_C \frac{dz}{(z + 2i)^2(z - 2i)^2} \\
&= 2\pi i \frac{d}{dz} \left[\frac{1}{(z + 2i)^2} \right]_{z=2i} \\
&= \frac{\pi}{16}
\end{aligned}
$$

Example 13

Evaluate the integral

$$\int_C \frac{z^2}{z + 1} dz$$

where C is the circle $|z - 2| = 1$.

Solution

The contour C is depicted in Fig. 2.10.

Since the function $z^2/(z + 1)$ is analytic everywhere inside the given contour C, the value of the integral $\int_C \{z^2/(z + 1)\} dz$ is zero.

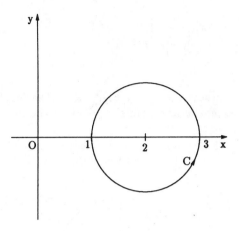

Fig. 2.10: A contour.

Example 14

Evaluate the following integral with the help of the Cauchy integral formula:
(a) $\int \{e^z/(z^2 - 3z)\}\, dz$ on $|z| = 1$
(b) $\int \{\sin z/(z^2 + 1)\}\, dz$ on $|z| = 2$

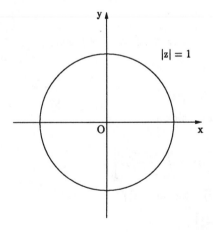

Fig. 2.11: A unit circle.

Solution

(a) The contour is a unit circle as shown in Fig. 2.11.
The function $e^z/z(z - 3)$ is analytic everywhere inside the unit circle except at $z = 0$.
Therefore,

$$\int \frac{e^z}{z(z-3)}dz = 2\pi i \left[\frac{e^z}{z-3}\right]_{z=0}$$
$$= -\frac{2\pi i}{3}$$

(b)   The closed contour is given by Fig. 2.12. The function $\sin z/(z^2+1)$ is analytic everywhere except at the points $z = \pm i$. Denoting the integral by I we obtain

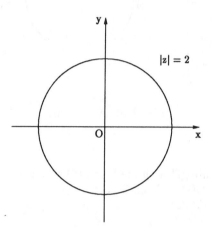

Fig. 2.12: A closed contour.

$$I = \int_C \frac{\sin z\, dz}{(z+i)(z-i)}$$
$$= \frac{1}{2i}\int_C \left[\frac{\sin z}{z-i} - \frac{\sin z}{z+i}\right]dz$$
$$= \frac{1}{2i}\left[\int_C \frac{\sin z}{z-i}dz - \int_C \frac{\sin z}{z+i}dz\right]$$
$$= \frac{1}{2i}\{2\pi i(\sin i) - 2\pi i \sin(-i)\}$$
$$= \pi\{\sin i + \sin i\}$$
$$= 2\pi \sin i$$
$$= 2\pi i \sinh 1$$

Example 15

Evaluate the integral by Cauchy integral formula

$$g(z_0) = \int_C \frac{z^3 + 2z}{(z - z_0)^3} dz$$

where C is a closed contour and the point z_0 is inside C (see Fig. 2.13).

Solution

Here $z = z_0$ is a triple pole which is inside the given contour C. Therefore by Cauchy integral formula we have

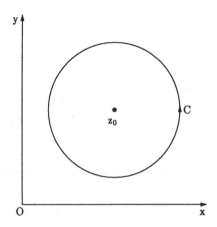

Fig. 2.13: A closed contour.

$$
\begin{aligned}
g(z_0) &= \int_C \frac{z^3 + 2z}{(z - z_0)^3} dz \\
&= (2\pi i) \times \lim_{z \to z_0} \frac{1}{2!} \frac{d^2}{dz^2} \left[(z - z_0)^3 \frac{z^3 + 2z}{(z - z_0)^3} \right] \\
&= (2\pi i) \times \lim_{z \to z_0} \frac{1}{2!} \frac{d^2}{dz^2} (z^3 + 2z) \\
&= (\pi i)(6z)_{z=z_0} = 6\pi i z_0
\end{aligned}
$$

If z_0 is outside the closed contour C, then

$$\int_C \frac{z^3 + 2z}{(z - z_0)^3} dz = g(z_0) = 0$$

Exercises

1. Find the value of

$$\int_C f(z)dz$$

   where $f(z) = x - y + 3y^2 i$ and C

   (a) is the line segment from $z = 0$ to $z = 1 + i$.

   (b) consists of two line segments, one from $z = 0$ to $z = i$ and the other from $z = i$ to $z = 1 + i$.

2. Evaluate

$$\int_0^{2+i} z^2 dz$$

   (a) Along the line $y = x/2$.

   (b) Along the real axis to 2 and then vertically to $2 + i$.

   (c) Along the imaginary axis to i and then horizontally to $2 + i$.

3. Evaluate $\displaystyle\int_0^{2+i} (\bar{z})^2 dz$ along each of the paths used in Exercise 2.

4. Evaluate $\displaystyle\int_C \frac{z+1}{z} dz$ where C is:

   (a) the semi-circle $z = e^{i\theta}$ $(0 \leq \theta \leq \pi)$;

   (b) the semi-circle $z = e^{i\theta}$ $(\pi \leq \theta \leq 2\pi)$;

   (c) the circle $z = e^{i\theta}$ $(-\pi \leq \theta \leq \pi)$.

5. Evaluate $\displaystyle\oint_C \frac{e^{4z}}{z^2+1} dz$ where C is the circle $|z| = 2$. What is the value of this integral if the path of integration is the circle $|z| = \frac{1}{2}$?

6. Evaluate $\displaystyle\oint_C \frac{z^2 + 3z + 2}{z + 1} dz$ where the contour C is given by

   (a) the circle $|z + 1| = 1$

   (b) the circle $|z + i| = 1$

   (c) the ellipse $\dfrac{x^2}{2} + \dfrac{y^2}{3} = 1$.

7. Show that if C_0 is the circle

$$z - z_0 = r_0 e^{i\theta} \quad (0 \le \theta \le 2\pi)$$

described counterclockwise, and $f(z)$ is continuous on C_0, then

$$\oint_{C_0} f(z)dz = i r_0 \int_0^{2\pi} f(z_0 + r_0 e^{i\theta})e^{i\theta}d\theta$$

Then obtain that

$$\oint_{C_0} \frac{dz}{z - z_0} = 2\pi i$$

$$\oint_{C_0} (z - z_0)^{n-1}dz = 0 \quad (n = \pm 1, \pm 2, \pm 3, \cdots)$$

8. Show that

$$\left| \oint_C \frac{\ln \sqrt{z}}{z^2}dz \right| < \pi \left(\frac{\ln R + \pi}{R} \right)$$

where the contour C is $|z| = R$, $R > 1$. Hence show that the value of the integral goes to zero as $R \to \infty$.

9. Determine the domain of analyticity of the function $f(z)$ and state why

$$\oint_C f(z)dz = 0$$

when the simple closed contour C is the unit circle $|z| = 1$ when

(a) $f(z) = \dfrac{z^2}{z - 2}$

(b) $f(z) = \tan z$

(c) $f(z) = \operatorname{sech} z$

(d) $f(z) = \ln(z + 2)$

(e) $f(z) = \dfrac{1}{z^2 + 4z + 13}$

(f) $f(z) = z e^{-z}$

10. Prove that

$$\oint_C z e^{-1/z}dz = \pi i$$

when C is $|z| = 1$. Explain why?

11. Using the substitution $z = e^{i\theta}$, show that

$$\int_0^{2\pi} \cos^{2n} \theta d\theta = \int_0^{2\pi} \sin^{2n} \vartheta d\theta = \frac{1}{2^{2n}} \frac{(2n)!}{(n!)^2} 2\pi \quad n = 1, 2, 3, \cdots$$

12. Show that

$$\oint_C \frac{e^{xz}}{z^{n+1}} dz = (2\pi i) \left(\frac{x^n}{n!} \right)$$

where C is any simple closed contour enclosing the origin.

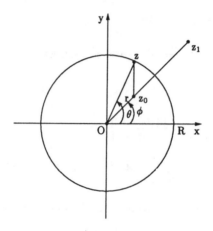

Fig. 2.14: Inverse point with respect to a circle.

13. Show that Cauchy's integral formula (see Fig. 2.14)

$$f(z_0) = \frac{1}{2\pi i} \oint_C \frac{f(z)}{z - z_0} dz$$

can be obtained as

$$f(r e^{i\phi}) = \frac{1}{2\pi i} \int_0^{2\pi} \frac{f(R e^{i\theta})}{R e^{i\theta} - r e^{i\phi}} (i R e^{i\theta} d\theta)$$

where

$$z = R e^{i\theta}$$
$$z_0 = r e^{i\phi}$$

Then prove that

$$\frac{1}{2\pi i} \int_0^{2\pi} \frac{f(R e^{i\theta})}{R e^{i\theta} - (R^2/r) e^{i\phi}} (i R e^{i\theta} d\theta) = 0$$

14. By subtracting the two integrals in Exercise 13 and equating the real parts in the resulting equation, obtain the Poisson integral for a circle:

$$u(r, \phi) = \frac{1}{2\pi} \int_0^{2\pi} \frac{(R^2 - r^2)u(R, \theta)}{R^2 - 2Rr\,\cos(\theta - \phi) + r^2} d\theta$$

Remark

The inverse of a nonzero point z_0 with respect to the circle (see Fig. 2.14) is the point z_1 lying on the same ray as z_0 and satisfying the condition $|z_1||z_0| = R^2$. Thus

$$z_1 = \frac{R^2}{r} e^{i\phi} = \frac{z\,\bar{z}}{\bar{z}_0}$$

Chapter 3

Series expansion

3.1 Power series

In this chapter we shall discuss the series representation of analytic functions in the form of power series. They are series of the form

$$\sum_{n=0}^{\infty} a_n(z - z_0)^n = a_0 + a_1(z - z_0) + \cdots + a_n(z - z_0)^n + \cdots \qquad (3.1)$$

where z_0, a_0 and a_n are complex constants and z may be any complex variable in a stated region.

We know from infinite series in real variable cases that the most useful test for convergence of infinite series is the ratio test. This test can be applied to series whose terms are complex. By ratio test, we have

$$\begin{aligned} \lim_{n \to \infty} \left| \frac{u_{n+1}}{u_n} \right| &= \lim_{n \to \infty} \left| \frac{a_{n+1}(z - z_0)^{n+1}}{a_n(z - z_0)^n} \right| \\ &= \lim_{n \to \infty} \left| \frac{a_{n+1}}{a_n}(z - z_0) \right| < 1 \end{aligned} \qquad (3.2)$$

for the series (3.1) to converge where u_n and u_{n+1} are the nth and $(n+1)$th terms of the series, respectively. Equation (3.2) can be written as

$$|z - z_0| < \lim_{n \to \infty} \left| \frac{a_n}{a_{n+1}} \right|$$

for the convergence of the series (3.1). Thus, if we define

$$r = \lim_{n \to \infty} \left| \frac{a_n}{a_{n+1}} \right|$$

as the radius of convergence, then the following theorem can be stated.

Theorem 3.1

Every power series $\sum_{n=0}^{\infty} a_n(z - z_0)^n$ has a radius of convergence r such that the series converges absolutely when $|z - z_0| < r$, and diverges when $|z - z_0| > r$ and no information can be obtained about the convergence or divergence of the series when $|z - z_0| = r$. That means, at the boundary of the circle $|z - z_0| = r$ of radius r and centre at z_0, no general statement can be made about the convergence. This boundary is formed on the circle of convergence (see Fig. 3.1).

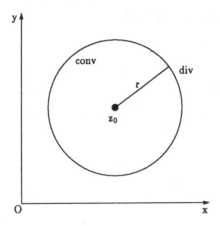

Fig. 3.1: Circle of convergence of a power series.

The following theorems are useful in connection with the power series and are cited without any formal proofs.

Theorem 3.2

A power series with non-zero convergence radius represents a continuous function with the circle of convergence.

Theorem 3.3

A power series can be integrated term by term within the circle of convergence, that is, if $r \neq 0$, and $f(z) = \sum_{n=0}^{\infty} a_n(z - z_0)^n$, $|z - z_0| < r$, then,

$$\int f(z)dz = \sum_{n=0}^{\infty} a_n \int_C (z - z_0)^n dz$$

$$= \sum_{n=0}^{\infty} a_n \frac{(z - z_0)^{n+1}}{n+1} + \text{const}$$

Theorem 3.4

A power series can be differentiated term by term, that is, if $r \neq 0$ and

$$f(z) = \sum_{n=0}^{\infty} a_n(z - z_0)^n, \qquad |z - z_0| < r$$

then

$$f'(z) = \sum_{n=1}^{\infty} n a_n(z - z_0)^{n-1}, \qquad |z - z_0| < r$$

Example 1

Find the region of convergence of the series

$$\frac{4 - z^2}{1^2} + \frac{(4 - z^2)^2}{2^2} + \frac{(4 - z^2)^3}{3^2} + \cdots$$

Solution

In this case, nth and $(n+1)$th terms can be written, respectively, as $u_n = (4 - z^2)^n/n^2$ and $u_{n+1} = (4 - z^2)^{n+1}/(n + 1)^2$.

The convergence of the series can be determined by the ratio test,

$$\lim_{n \to \infty} \left| \frac{u_{n+1}}{u_n} \right| < 1$$

Therefore the given series will be convergent if

$$\lim_{n \to \infty} \left| \frac{(4 - z^2)^{n+1}/(n + 1)^2}{(4 - z^2)^n/n^2} \right| < 1$$

$$\lim_{n \to \infty} \left| \frac{(4 - z^2)n^2}{(n + 1)^2} \right| < 1$$

After reduction we obtain

$$|(4 - z^2)| < \lim_{n \to \infty} \left(\frac{n + 1}{n} \right)^2$$

$$|4 - z^2| < 1$$

This can be subsequently reduced to

$$|4 - (x^2 - y^2 + 2ixy)| < 1$$
$$|(4 - x^2 + y^2) - 2ixy| < 1$$
$$(4 - x^2 + y^2)^2 + 4x^2y^2 < 1$$
$$(x^2 - y^2 - 4)^2 + 4x^2y^2 < 1$$

Hence the given series is convergent inside the lemniscate the algebraic equation of which is given by

$$(x^2 - y^2 - 4)^2 + 4x^2y^2 = 1$$

Example 2

Determine the region of convergence of the series

$$\frac{1}{z(z+1)} + \frac{1}{(z+1)(z+2)} + \frac{1}{(z+2)(z+3)} + \cdots$$

Solution

In this case, nth and $(n+1)$th terms are given by $u_n = 1/\{(z+n-1)(z+n)\}$ and $u_{n+1} = 1/\{(z+n)(z+n+1)\}$.

The convergence of the series is determined by the ratio test,

$$\lim_{n \to \infty} \left| \frac{u_{n+1}}{u_n} \right| < 1$$

Therefore the given series is convergent if

$$\lim_{n \to \infty} \left| \frac{z+n-1}{z+n+1} \right| < 1$$

or

$$\lim_{n \to \infty} \left| 1 - \frac{2}{z+n+1} \right| < 1$$

$$|z| < \infty$$

The series converges for all values of z, except for $z = 0, -1, -2, \cdots$. Because it is evident from the given series, we have

$$\left(\frac{1}{z} - \frac{1}{z+1} \right) + \left(\frac{1}{z+1} - \frac{1}{z+2} \right) + \cdots \left(\frac{1}{z+n} - \frac{1}{z+n+1} \right) + \cdots$$

the sum of which is $1/z$.

Exercises

Determine the region of convergence and the sum of each of the following series:

1. $1 + z + z^2 + \cdots$

2. $1 + (z-1) + (z-1)^2 + \cdots$

3. $1 - (z-1) + (z-1)^2 - \cdots$

4. $1 - 2(z-2) + 3(z-2)^2 - 4(z-2)^3 + \cdots$

5. $\dfrac{1}{2} \left(\dfrac{z+2}{z-2} \right) + \dfrac{1}{2^2} \left(\dfrac{z+2}{z-2} \right)^2 + \dfrac{1}{2^3} \left(\dfrac{z+2}{z-2} \right)^3 + \cdots$

6. $\dfrac{1}{2(z-i)} + \dfrac{1}{2^2(z-i)^2} + \dfrac{1}{2^3(z-i)^3} + \cdots$

7. Show that

$$\sum_{n=1}^{\infty} z^n = \frac{z}{1-z}$$

where z is a complex number such that $|z| < 1$.

8. In the formula of Exercise 7 write $z = r\,e^{i\theta}$, where $0 < r < 1$. Show that

$$\sum_{n=1}^{\infty} r^n \cos n\theta = \frac{r\cos\theta - r^2}{1 - 2r\cos\theta + r^2}$$

$$\sum_{n=1}^{\infty} r^n \sin n\theta = \frac{r\sin\theta}{1 - 2r\cos\theta + r^2}$$

9. Find the region of convergence of the series

$$\frac{z^2-1}{1^2} + \frac{(z^2-1)^2}{2^2} + \frac{(z^2-1)^3}{2^3} + \cdots$$

10. Determine the region of convergence and the sum of the series

$$\frac{1}{z(z+1)} + \frac{1}{(z+1)(z+2)} + \frac{1}{(z+2)(z+3)} + \cdots$$

Find the region of convergence and the sum of each of the following series:

11. $\displaystyle\sum_{n=0}^{\infty} e^{-nz}$

12. $\displaystyle\sum_{n=0}^{\infty} e^{-n(z^2+1)}$

13. $\displaystyle\sum_{n=0}^{\infty} e^{n/z}$

14. $\displaystyle\sum_{n=1}^{\infty} n\,z\,e^{-nz^2}$

15. Show in two ways that the sequence

$$z_n = -2 + i\frac{(-1)^n}{n}, \quad n = 1, 2, \cdots$$

converges.

16. For what values of z does the series $\displaystyle\sum_{n=1}^{\infty} \frac{1}{(z^2+1)^n}$ converge? Determine its sum.

3.2 Taylor series

We shall be concerned here with Taylor's theorem which has many applications in practical problems.

Theorem 3.5

If f be analytic everywhere inside a circle C_0 with centre at z_0 and radius r_0, then at each point inside C_0

$$
\begin{aligned}
f(z) &= f(z_0) + (z - z_0)f'(z_0) + \frac{(z - z_0)^2}{2!}f''(z_0) + \cdots \\
&\quad + \frac{(z - z_0)^{n-1}}{(n-1)!}f^{(n-1)}(z_0) + R_n(z)
\end{aligned}
\tag{3.3}
$$

where

$$
R_n(z) = \frac{(z - z_0)^n}{2\pi i} \int_{C_1} \frac{f(\xi)d\xi}{(\xi - z)(\xi - z_0)^n}
$$

in which C_1 is a circle inside C_0 as shown in Fig. 3.2.

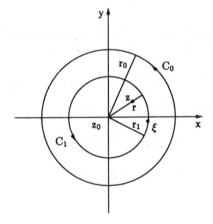

Fig. 3.2: C_0 and C_1 two circles used to establish Taylor's theorem.

Proof

To prove the theorem, we let z be any fixed point inside the circle C_0 and write $|z - z_0| = r$; thus $r < r_0$. We also let ξ be any point lying on a circle C_1, centred at z_0 and with radius $r < r_1 < r_0$; and thus $|\xi - z_0| = r_1$. By Cauchy's integral formula, we can write

$$f(z) = \frac{1}{2\pi i} \int_{C_1} \frac{f(\xi)d\xi}{(\xi - z)}$$

$$= \frac{1}{2\pi i} \int_{C_1} \frac{f(\xi)d\xi}{(\xi - z_0) - (z - z_0)}$$

$$= \frac{1}{2\pi i} \int_{C_1} \frac{f(\xi)d\xi}{(\xi - z_0)} \frac{1}{1 - (z - z_0)/(\xi - z_0)} \tag{3.4}$$

Then applying the identity $1/(1 - v) = 1 + v + v^2 + \cdots + v^{n-1} + v^n/(1 - v)$ we have,

$$\frac{1}{1 - (z - z_0)/(\xi - z_0)} = 1 + \frac{z - z_0}{\xi - z_0} + \left(\frac{z - z_0}{\xi - z_0}\right)^2 + \cdots + \left(\frac{z - z_0}{\xi - z_0}\right)^{n-1}$$

$$+ \frac{1}{1 - (z - z_0)/(\xi - z_0)} \left(\frac{z - z_0}{\xi - z_0}\right)^n \tag{3.5}$$

Therefore from (3.4), we have

$$f(z) = \frac{1}{2\pi i} \int_{C_1} \frac{f(\xi)}{(\xi - z_0)} \left[1 + \left(\frac{z - z_0}{\xi - z_0}\right) + \left(\frac{z - z_0}{\xi - z_0}\right)^2 + \cdots + \left(\frac{z - z_0}{\xi - z_0}\right)^{n-1}\right.$$

$$\left. + \frac{(\frac{z-z_0}{\xi-z_0})^n}{1 - (z - z_0)/(\xi - z_0)}\right] d\xi$$

$$= \frac{1}{2\pi i} \int_{C_1} \frac{f(\xi)d\xi}{\xi - z_0} + \frac{(z - z_0)}{2\pi i} \int_{C_1} \frac{f(\xi)d\xi}{(\xi - z_0)^2} + \frac{(z - z_0)^2}{2\pi i} \int_{C_1} \frac{f(\xi)d\xi}{(\xi - z_0)^3} + \cdots$$

$$+ \frac{(z - z_0)^{n-1}}{2\pi i} \int_{C_1} \frac{f(\xi)d\xi}{(\xi - z_0)^n} + \frac{(z - z_0)^n}{2\pi i} \int_{C_1} \frac{f(\xi)d\xi}{(\xi - z)(\xi - z_0)^n}$$

In view of Cauchy's integral formula, we have

$$f(z) = f(z_0) + (z - z_0)f'(z_0) + \frac{(z - z_0)^2}{2!} f''(z_0) + \cdots$$

$$+ \frac{(z - z_0)^{n-1}}{(n - 1)!} f^{(n-1)}(z_0) + R_n(z) \tag{3.6}$$

where

$$R_n(z) = \frac{(z - z_0)^n}{2\pi i} \int_{C_1} \frac{f(\xi)d\xi}{(\xi - z)(\xi - z_0)^n} \tag{3.7}$$

Now with reference to Fig. 3.2, we know that $|z - z_0| = r$ and $|\xi - z_0| = r_1$.

Also we note that $|\xi - z| \geq |\xi - z_0| - |z - z_0|$ and $|\xi - z| \geq r_1 - r$.

Hence it follows from (3.7) that when M denotes the maximum value of $f(\xi)$ on C_1,

$$|R_n(z)| \leq \frac{r^n}{2\pi} \frac{M2\pi r_1}{(r_1 - r)r_1^n}$$

$$= \frac{Mr_1}{r_1 - r}(\frac{r}{r_1})^n$$

But $(r/r_1) < 1$, and therefore

$$\lim_{n \to \infty} |R_n(z)| = 0$$

Hence, for each point z interior to C_0, the limit as $n \to \infty$ of the right-hand side is $f(z)$. Thus, if f is analytic inside a circle centred at z_0 with radius r_0, then $f(z)$ is represented by a Taylor series:

$$f(z) = f(z_0) + \sum_{n=1}^{\infty} \frac{(z - z_0)^n}{n!} f^{(n)}(z_0), \text{ when } |z - z_0| < r_0 \qquad (3.8)$$

If $z_0 = 0$, then this for a MacLaurin series gives:

$$f(z) = f(0) + \sum_{n=1}^{\infty} \frac{z^n}{n!} f^{(n)}(0), \text{ when } |z| < r_0 \qquad (3.9)$$

3.2.1 Taylor series of elementary functions

From the Taylor series presented above it may be considered that if f is analytic at all points within a circle centred at z_0, convergence of Taylor series about z_0 to $f(z)$ for each point z with that circle is guaranteed. In fact, Taylor's theorem says that the series converges to $f(z)$ within the circle about z_0 whose radius is the distance from z_0 to the point z, at which f is not analytic. In the following we shall present MacLaurin series expansions of some elementary functions.

MacLaurin series expansion of e^z can be written as

$$e^z = 1 + z + \frac{z^2}{2!} + \frac{z^3}{3!} + \cdots$$

$$= 1 + \sum_{n=1}^{\infty} \frac{z^n}{n!} \qquad \text{when } |z| < \infty \qquad (3.10)$$

$$\sin z = \sum_{n=1}^{\infty} (-1)^{n+1} \frac{z^{2n-1}}{(2n - 1)!} \qquad \text{when } |z| < \infty \qquad (3.11)$$

$$\cos z = 1 + \sum_{n=1}^{\infty} (-1)^n \frac{z^{2n}}{(2n)!} \qquad \text{when } |z| < \infty \qquad (3.12)$$

$$\sinh z = \sum_{n=1}^{\infty} \frac{z^{2n-1}}{(2n - 1)!} \qquad \text{when } |z| < \infty \qquad (3.13)$$

$$\cosh z = 1 + \sum_{n=1}^{\infty} \frac{z^{2n}}{(2n)!} \qquad \text{when } |z| < \infty \qquad (3.14)$$

MacLaurin series that is readily obtained is

$$\frac{1}{1+z} = \sum_{n=0}^{\infty}(-1)^n z^n, \qquad \text{when } |z| < 1$$

Also expansion of $1/z$ into a Taylor series about the point $z = 1$, is

$$\begin{aligned} \frac{1}{z} &= \frac{1}{[1+(z-1)]} \\ &= [1+(z-1)]^{-1} \\ &= \sum_{n=0}^{\infty}(-1)^n(z-1)^n \qquad \text{when } |z-1| < 1 \end{aligned}$$

Now we shall present some problems below concerning the Taylor and MacLaurin series.

Example 1

Show that

$$e^z = e + e\sum_{n=1}^{\infty}\frac{(z-1)^n}{n!} \qquad \text{when } |z-1| < \infty$$

Proof

In this case, put $z = 1 + u$, then

$$\begin{aligned} e^z &= e^{1+u} \\ &= e \cdot e^u \\ &= e\left[1 + \sum_{n=1}^{\infty}\frac{u^n}{n!}\right] \\ &= e + e\sum_{n=1}^{\infty}\frac{(z-1)^n}{n!} \qquad \text{when } |z-1| < \infty \end{aligned}$$

as asserted.

Example 2

Show that

$$\frac{1}{z^2} = 1 + \sum_{n=1}^{\infty}(n+1)(z+1)^n, \qquad \text{when } |z+1| < 1$$

Proof

We find that

$$\frac{1}{z} = \frac{1}{1+z-1}$$

$$= \frac{-1}{1-(z+1)} = -[1-(z+1)]^{-1}$$

$$= -[1+(z+1)+(z+1)^2+(z+1)^3+\cdots] \qquad \text{when } |z+1|<1$$

Now differentiating left-hand side and right-hand side with respect to z, we have, (which is allowed for convergence series)

$$\frac{1}{z^2} = [0+1+2(z+1)+z(z+1)^2+\cdots+(n+1)(z+1)^n+\cdots]$$

$$= 1+\sum_{n=1}^{\infty}(n+1)(z+1)^n \qquad \text{when } |z+1|<1$$

as asserted.

Example 3

Show that

$$\frac{1}{z^2} = \frac{1}{4}+\frac{1}{4}\sum_{n=1}^{\infty}(-1)^n(n+1)\left(\frac{z-2}{2}\right)^n, \qquad \text{when } |z-2|<2$$

Proof

We can write

$$\frac{1}{z^2} = \frac{1}{(z-2+2)^2}$$

$$= \frac{1}{4\{1+(z-2)/2\}^2}$$

$$= \frac{1}{4}\left\{1+\left(\frac{z-2}{2}\right)\right\}^{-2}$$

We know that

$$(1+y)^{-n} = 1+(-n)y+\frac{(-n)(-n-1)}{2!}y^2+\cdots \qquad \text{when } |y|<1$$

Therefore,

$$(1+y)^{-2} = 1-2y+3y^2-4y^3+\cdots$$

$$= 1+\sum_{n=1}^{\infty}(-1)^n(n+1)y^n$$

Here,

$$\frac{1}{z^2} = \frac{1}{4}\left[1 + \sum_{n=1}^{\infty}(-1)^n(n+1)\left(\frac{z-2}{2}\right)^n\right]$$

$$= \frac{1}{4} + \frac{1}{4}\sum_{n=1}^{\infty}(-1)^n(n+1)\left(\frac{z-2}{2}\right)^n \qquad \text{when } |z-2| < 2$$

as asserted.

Example 4

Prove that when $0 < |z| < 4$,

$$\frac{1}{4z - z^2} = \sum_{n=0}^{\infty}\frac{z^{n-1}}{4^{n+1}}.$$

Proof

In this case,

$$\frac{1}{4z - z^2} = \frac{1}{4z}\frac{1}{(1 - z/4)}$$

$$= \frac{1}{4z}\left(1 + \frac{z}{4} + \frac{z^2}{16} + \cdots\right) \qquad \text{when } 0 < |z| < 4$$

$$= \frac{1}{4z} + \frac{1}{4^2} + \frac{z}{4^3} + \frac{z^2}{4^4} + \cdots + \frac{z^{n-1}}{4^{n+1}} + \cdots$$

$$= \sum_{n=0}^{\infty}\frac{z^{n-1}}{4^{n+1}} \qquad \text{when } 0 < |z| < 4$$

as asserted.

Exercises

Expand each of the following functions in a Taylor series about the indicated points. Determine the radius of convergence of each series:

1. $f(z) = \dfrac{z+1}{z-1}, \quad z = 0, z = -1$

2. $f(z) = \dfrac{1}{z(z+1)}, \quad z = 0, z = -1$

3. $f(z) = \cos z, \quad z = \dfrac{\pi}{2}$

4. $f(z) = \cosh z, \quad z = i\pi$

5. $f(z) = \tan^{-1} z, \quad z = 1$

6. $f(z) = \dfrac{1}{1 - e^z}, \quad z = 4i$

Show that

7. $\dfrac{1}{z^2} = \displaystyle\sum_{n=0}^{\infty} (n+1)(z+1)^n \quad$ for $|z+1| < 1$

8. $\dfrac{1}{z^2} = \dfrac{1}{4} \displaystyle\sum_{n=0}^{\infty} (-1)^n (n+1) \left(\dfrac{z-2}{2} \right) \quad$ for $|z-2| < 2$

9. $\dfrac{1}{2z - z^2} = \displaystyle\sum_{n=0}^{\infty} \dfrac{z^{n-1}}{2^{n+1}} \quad$ for $0 < |z| < 2$

10. $\dfrac{\sin(z^2)}{z^3} = \dfrac{1}{z} - \dfrac{z^3}{z!} + \dfrac{z^7}{5!} - \dfrac{z^{11}}{7!} + \cdots \quad$ for $|z| < \infty$

11. Show that

$$\dfrac{z}{(z-1)(z-3)} = \dfrac{-1}{2(z-1)} - 3 \sum_{n=1}^{\infty} \dfrac{(z-1)^{n-1}}{2^{n+1}} \quad \text{when } 0 < |z-1| < 2$$

12. Prove that

$$\dfrac{1}{1+z^2} = 1 - z^2 + z^4 - z^6 + \cdots \quad \text{for } |z| \leq 1$$

13. Using the result in Exercise 12,

(a) prove that

$$\tan^{-1} z = \int_0^z \dfrac{1}{1+z^2} dz$$
$$= z - \dfrac{z^3}{3} + \dfrac{z^5}{5} - \dfrac{z^7}{7} + \cdots$$

(b) prove that

$$\dfrac{\pi}{4} = 1 - \dfrac{1}{3} + \dfrac{1}{5} - \dfrac{1}{7} + \cdots$$

14. Prove that

$$\dfrac{1}{\sqrt{1-z^2}} = 1 + \dfrac{1}{2} z^2 + \dfrac{1\cdot 3}{2!4} z^4 + \cdots \quad \text{for } |z| < 1$$

15. Using the result in Exercise 14, show that

$$\sin^{-1} z = \int_0^z \dfrac{dz}{\sqrt{1-z^2}} = z + \dfrac{z^3}{2\cdot 3} + \dfrac{1\cdot 3}{2!4\cdot 5} z^5 + \cdots \quad |z| < 1$$

3.3   Laurent series

In many practical applications, it is necessary to expand a function about a point at which or in the neighbourhood of which the function is not analytic. In this situation, it is obvious that the Taylor series expansion is not applicable and a Laurent series expansion is needed. We state the following theorem known as Laurent's Theorem.

Theorem 3.6

If C_1 and C_2 are the two concentric circles centred at z_0 and with the radii r_1 and r_2, respectively, where $r_2 < r_1$ as shown in Fig. 3.3. If f is analytic on C_1 and C_2 and throughout the annular region between these two circles, then at each point z in that domain $f(z)$ is represented by the expansion

$$f(z) = \sum_{n=0}^{\infty} a_n(z - z_0)^n + \sum_{n=1}^{\infty} \frac{b_n}{(z - z_0)^n} \tag{3.15}$$

where

$$a_n = \frac{1}{2\pi i} \int_{C_1} \frac{f(\xi)d\xi}{(\xi - z_0)^{n+1}} \quad (n = 0, 1, 2, \dots) \tag{3.16}$$

$$b_n = \frac{1}{2\pi i} \int_{C_2} \frac{f(\xi)d\xi}{(\xi - z_0)^{-n+1}} \quad (n = 1, 2, \dots) \tag{3.17}$$

each path of integration being taken anticlockwise. Series (3.15) is called a Laurent series.

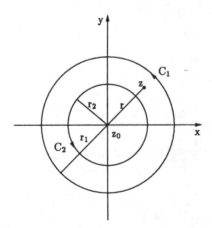

Fig. 3.3: Contours C_1 and C_2 to describe the Laurent series.

Proof

Consider a point z inside the annulus of C_1 and C_2. Then, by Cauchy's integral formula, we can write

$$f(z) = \frac{1}{2\pi i} \int_{C_1} \frac{f(\xi)d\xi}{(\xi - z)} - \frac{1}{2\pi i} \int_{C_2} \frac{f(\xi)d\xi}{(\xi - z)}$$

$$= \frac{1}{2\pi i} \int_{C_1} \frac{f(\xi)d\xi}{(\xi - z_0) - (z - z_0)} + \frac{1}{2\pi i} \int_{C_2} \frac{f(\xi)d\xi}{(z - z_0) - (\xi - z_0)}$$

$$= \frac{1}{2\pi i} \int_{C_1} \frac{f(\xi)d\xi}{(\xi - z_0)\{1 - (z - z_0)/(\xi - z_0)\}}$$

$$+ \frac{1}{2\pi i} \int_{C_2} \frac{f(\xi)d\xi}{(z - z_0)\{1 - (\xi - z_0)/(z - z_0)\}}$$

The point ξ is a dummy variable, when we consider C_1 integral, ξ is on C_1 and in the case of second integral C_2, ξ is considered to be on C_2.

Now in each of these integrals let us apply the identity:

$$\frac{1}{1 - v} = 1 + v + v^2 + \cdots + \frac{v^n}{1 - v}$$

Then we have

$$f(z) = \frac{1}{2\pi i} \int_{C_1} \frac{f(\xi)}{(\xi - z_0)} \left[1 + \frac{z - z_0}{\xi - z_0} + \left(\frac{z - z_0}{\xi - z_0} \right)^2 \right.$$

$$\left. + \cdots + \frac{1}{1 - (\frac{z - z_0}{\xi - z_0})} \left(\frac{z - z_0}{\xi - z_0} \right)^n \right] d\xi$$

$$+ \frac{1}{2\pi i} \int_{C_2} \frac{f(\xi)}{(z - z_0)} \left[1 + (\frac{\xi - z_0}{z - z_0}) + (\frac{\xi - z_0}{z - z_0})^2 \right.$$

$$\left. + \cdots + \frac{1}{1 - \frac{\xi - z_0}{z - z_0}} \left(\frac{\xi - z_0}{z - z_0} \right)^n \right] d\xi$$

$$= \frac{1}{2\pi i} \int_{C_1} \frac{f(\xi)d\xi}{(\xi - z_0)} + \frac{z - z_0}{2\pi i} \int_{C_1} \frac{f(\xi)d\xi}{(\xi - z_0)^2} + \frac{(z - z_0)^2}{2\pi i} \int_{C_1} \frac{f(\xi)d\xi}{(\xi - z_0)^3}$$

$$+ \cdots + \frac{(z - z_0)^{n-1}}{2\pi i} \int_{C_1} \frac{f(\xi)d\xi}{(\xi - z_0)^n} + R_n(z)$$

$$+ \frac{1}{2\pi i (z - z_0)} \int_{C_2} f(\xi)d\xi + \frac{1}{2\pi i (z - z_0)^2} \int_{C_2} f(\xi)(\xi - z_0)d\xi$$

$$+ \frac{1}{2\pi i (z - z_0)^3} \int_{C_2} (\xi - z_0)^2 f(\xi)d\xi + \cdots +$$

$$+ \frac{1}{2\pi i (z - z_0)^n} \int_{C_2} (\xi - z_0)^{n-1} f(\xi)d\xi + Q_n(z) \tag{3.18}$$

where

$$R_n = \frac{(z - z_0)^n}{2\pi i} \int_{C_1} \frac{f(\xi)d\xi}{(\xi - z)(\xi - z_0)^n} \tag{3.19}$$

$$Q_n = \frac{1}{2\pi i(z - z_0)^n} \int_{C_2} \frac{(\xi - z_0)^n f(\xi)d\xi}{(z - \xi)} \tag{3.20}$$

The theorem will be established provided we can show that

$$\lim_{n \to \infty} R_n = 0 \tag{3.21}$$

$$\lim_{n \to \infty} Q_n = 0 \tag{3.22}$$

To prove (3.21) we notice that the dummy variable point ξ cannot be on the circle C_2; and thus, $|z - z_0| = r$, $|\xi - z_0| = r_1$ and therefore we have $|\xi - z| = |(\xi - z_0) - (z - z_0)| > r_1 - r$. We know that $|f(\xi)| \leq M$ and $r < r_1$ and hence

$$
\begin{aligned}
|R_n| &= \left| \frac{(z - z_0)^n}{2\pi i} \int_{C_1} \frac{f(\xi)d\xi}{(\xi - z)(\xi - z_0)^n} \right| \\
&\leq \frac{r^n}{2\pi} \frac{M 2\pi r_1}{r_1^n(r_1 - r)} \\
&= \left(\frac{r}{r_1}\right)^n \frac{M r_1}{(r_1 - r)}
\end{aligned}
$$

Since $(r/r_1) < 1$ and hence as $n \to \infty$, $\lim_{n \to \infty} |R_n| \to 0$.

To prove (3.22), we notice that the dummy variable point ξ must be on the circle C_2, and thus $|z - z_0| = r$, $|\xi - z_0| = r_2$ and therefore $|z - \xi| = |(z - z_0) - (\xi - z_0)| > r - r_2$. We know $|f(\xi)| \leq M$ and $r_2 < r$ and hence

$$
\begin{aligned}
|Q_n| &= \left| \frac{1}{2\pi i(z - z_0)^n} \int_{C_2} \frac{(\xi - z_0)^n f(\xi)d\xi}{(z - \xi)} \right| \\
&\leq \frac{1}{2\pi r^n} \cdot \frac{r_2^n M}{r - r_2} \cdot 2\pi r_2 \\
&= \left(\frac{r_2}{r}\right)^n \left(\frac{M r_2}{r - r_2}\right)
\end{aligned}
$$

Since $(r_2/r) < 1$ and hence as $n \to \infty$, $\lim_{n \to \infty} |Q_n| \to 0$.

Thus $f(z)$ can be expanded by a Laurent series as

$$
\begin{aligned}
f(z) &= a_0 + a_1(z - z_0) + a_2(z - z_0)^2 + \cdots + a_n(z - z_0)^n + \\
&\quad + \frac{b_1}{z - z_0} + \frac{b_2}{(z - z_0)^2} + \cdots + \frac{b_n}{(z - z_0)^n} + \cdots \\
&= \sum_{n=0}^{\infty} a_n(z - z_0)^n + \sum_{n=1}^{\infty} \frac{b_n}{(z - z_0)^n}
\end{aligned}
$$

where a_n and b_n are given by (3.16) and (3.17) respectively. Hence this is the required proof.

Remark

It is to be noted here that if $f(z)$ is an analytic function at every point within and on the contour C_1 except at $z = z_0$ itself, the radius r_2 may be taken arbitrarily small. The Laurent expansion (3.15) is then valid for the domain $0 < |z - z_0| < r_1$. If $f(z)$ is analytic at all points within and on C_2, then the function $\frac{f(z)}{(z-z^{-n+1})}$ is analytic within and on C_2 since $-n + 1 \leq 0$ always. Therefore the integral in formula (3.17) has the value zero and expansion (3.15) reduces to a Taylor series.

It is worth mentioning that since the function $\frac{f(z)}{(z-z^{n+1})}$ and $\frac{f(z)}{(z-z^{-n+1})}$ are analytic throughout the annular region $r_2 \leq |z - z_0| \leq r_1$, any simple closed contour C around the annulus in the positive direction can be used as a path of integration in place of circular paths C_1 and C_2.

Example 1

If $f(z) = (z + 1)/(z - 1)$, find its

(a) MacLaurin series expansion and give the region of validity for that representation.

(b) Its Laurent series for the domain $|z| > 1$.

Solution

$$f(z) = \frac{z+1}{z-1}$$

which is not analytic at $z = 1$. Now rewriting $f(z)$ in the following way,

$$
\begin{aligned}
f(z) &= -\frac{z+1}{1-z} \\
&= -(z+1)(1-z)^{-1}
\end{aligned}
$$

This can be expanded in power series if $|z| < 1$. Thus we have

$$
\begin{aligned}
f(z) &= -(1+z)[1 + z + z^2 + z^3 + \cdots] \\
&= -1 - (z + z^2 + z^3 + \cdots) - (z + z^2 + \cdots) \\
&= -1 - 2\sum_{n=1}^{\infty} z^n, \qquad |z| < 1
\end{aligned}
$$

Since this series in powers of z converges to $f(z)$ when $|z| < 1$, it is the MacLaurin series for $f(z)$.

(b) We have to rewrite the function $f(z)$ as

$$
\begin{aligned}
f(z) &= \frac{1}{z}\frac{z+1}{1-1/z} \\
&= \left(\frac{z+1}{z}\right)\left(1 - \frac{1}{z}\right)^{-1}
\end{aligned}
$$

This can be expanded in power series if $|1/z| < 1$. Thus we have

$$
\begin{aligned}
f(z) &= \left(\frac{z+1}{z}\right)\left[1 + \frac{1}{z} + \frac{1}{z^2} + \cdots\right] \\
&= (z+1)\left[\frac{1}{z} + \frac{1}{z^2} + \frac{1}{z^3} + \cdots\right] \\
&= 1 + 2\sum_{n=1}^{\infty} z^{-n} \qquad \text{for } |z| > 1
\end{aligned}
$$

which is the Laurent series expansion.

Example 2

Obtain the expansion of the function $(z-1)/z^2$ into:
   (a) its Taylor series in powers of $(z-1)$ and give the region of validity; and
   (b) its Laurent's series for the domain $|z-1| > 1$.

Solution

(a) The function can be rewritten as

$$
f(z) = \frac{z-1}{(z-1+1)^2} = (z-1)[1 + (z-1)]^{-2}
$$

This can be expanded in power series provided $|z-1| < 1$.
So

$$
\begin{aligned}
f(z) &= (z-1)[1 - 2(z-1) + 3(z-1)^2 - 4(z-1)^3 + \cdots] \\
&= \sum_{n=1}^{\infty} (-1)^{n+1} n (z-1)^n
\end{aligned}
$$

which is Taylor's expansion about $z = 1$, i.e., in powers of $(z-1)$.
   (b) To find the Laurent series, we rearrange the terms in $f(z)$ as follows:

$$
f(z) = \frac{(z-1)}{(z-1)^2\{1 + 1/(z-1)\}^2} = \left(\frac{1}{z-1}\right)\left\{1 + \frac{1}{z-1}\right\}^{-2}
$$

This can be expanded in power series provided $|z-1| > 1$.
Thus expanding

$$
\begin{aligned}
f(z) &= \left(\frac{1}{z-1}\right)\left[1 - \frac{2}{z-1} + \frac{3}{(z-1)^2} - \cdots\right] \\
&= \frac{1}{z-1} - \frac{2}{(z-1)^2} + \frac{3}{(z-1)^3} - \cdots \\
&= \sum_{n=1}^{\infty} (-1)^{n+1} \frac{n}{(z-1)^n} \qquad |z-1| > 1
\end{aligned}
$$

which is the Laurent series.

Example 3

Obtain two Laurent series expansions in powers of z for the function

$$f(z) = \frac{1}{z^2(1-z)}$$

and give the regions of validity.

Solution

(i) The given function can be rearranged as follows:

$$f(z) = \frac{1}{z^2(1-z)} = \frac{1}{z^2}(1-z)^{-1}$$

This can be expanded in power series provided $|z| < 1$. Therefore,

$$
\begin{aligned}
f(z) &= \frac{1}{z^2}[1 + z + z^2 + z^3 + \cdots], \qquad |z| < 1 \\
&= \frac{1}{z^2} + \frac{1}{z} + 1 + \sum_{n=1}^{\infty} z^n \\
&= \sum_{n=0}^{\infty} z^{n-2}
\end{aligned}
$$

which is a Laurent series expansion valid in the range $0 < |z| < 1$.

(ii) Rearranging the terms we have

$$
\begin{aligned}
f(z) &= \frac{1}{z^2(1-z)} \\
&= \frac{1}{-z^3(1-\frac{1}{z})} \\
&= -\frac{1}{z^3}(1-\frac{1}{z})^{-1}
\end{aligned}
$$

This can be expanded in power series provided $|z| > 1$. Therefore,

$$
\begin{aligned}
f(z) &= -\frac{1}{z^3}\left[1 + \frac{1}{z} + \frac{1}{z^2} + \cdots\right] \\
&= -\sum_{n=0}^{\infty} z^{-n-3}
\end{aligned}
$$

which is a Laurent series which is valid in the range $|z| > 1$.

Example 4

Write the Laurent series expansion of the function $f(z) = 1/(z - k)$ for the domain $|z| > |k|$, where k is real and $-1 < k < 1$. Then obtain the formulas by putting $z = e^{i\theta}$:

$$\sum_{n=1}^{\infty} k^n \cos n\theta = \frac{k \cos\theta - k^2}{1 - 2k \cos\theta + k^2}$$

$$\sum_{n=1}^{\infty} k^n \sin n\theta = \frac{k \sin\theta}{1 - 2k \cos\theta + k^2}$$

Solution

Given that

$$f(z) = \frac{1}{z - k} = \frac{1}{z}\left(1 - \frac{k}{z}\right)^{-1}$$

This can be expanded in power series provided $|k/z| < 1$, i.e. $|z| > |k|$. Therefore, expanding

$$\begin{aligned}
f(z) &= \frac{1}{z}\left[1 + \frac{k}{z} + \frac{k^2}{z^2} + \cdots\right] \\
&= \sum_{n=0}^{\infty} \frac{k^n}{z^{n+1}} \qquad |z| > |k| \\
&= \frac{1}{z}\sum_{n=0}^{\infty} \frac{k^n}{z^n}
\end{aligned}$$

Thus, we have

$$\frac{z}{z - k} = \sum_{n=0}^{\infty} \left(\frac{k}{z}\right)^n$$

which can be rewritten as

$$1 + \frac{k}{z - k} = 1 + \sum_{n=1}^{\infty} \frac{k^n}{z^n}$$

and hence we have

$$\frac{k}{z - k} = \sum_{n=1}^{\infty} \frac{k^n}{z^n}$$

Then substituting $z = e^{i\theta} = \cos\theta + i \sin\theta$ in the last expression yields

$$\frac{k}{(\cos\theta - k) + i\ \sin\theta} = \sum_{n=1}^{\infty} k^n[\cos\ n\theta - i\ \sin n\theta]$$

$$\frac{k[(\cos\theta - k) - i\ \sin\theta]}{(\cos\theta - k)^2 + \sin^2\theta} = \sum_{n=1}^{\infty} k^n[\cos n\theta - i\ \sin n\theta]$$

Equating real and imaginary parts:

$$\sum_{n=1}^{\infty} k^n \cos n\theta = \frac{k\ \cos\theta - k^2}{1 - 2k\ \cos\theta + k^2}$$

$$\sum_{n=1}^{\infty} k^n \sin n\theta = \frac{k\ \sin\theta}{1 - 2k\ \cos\theta + k^2}$$

Hence they are the required results.

Exercises

1. Expand $f(z) = \dfrac{1}{z-2}$ in a Laurent series valid for $|z| > 2$ and $|z| < 2$

2. Expand $f(z) = \dfrac{1}{(z-2)(z-3)}$:

   (a) for $|z| < 2$

   (b) for $2 < |z| < 3$

   (c) for $|z| > 3$

   (d) for $0 < |z - 2| < 2$

   (e) for $|z - 2| > 2$

   (f) for $0 < |z - 3| < 2$

   (g) for $|z - 3| > 2$

3. Expand $f(z) = \dfrac{2z+1}{z^2-1}$ in two different Laurent expansions around $z = 1$ stating clearly the region of convergence in each case.

4. Find a Laurent series expansion of $f(z) = e^{z/(z-2)}$ about $z = 2$ and state the region of convergence of this series.

5. Find the value of $\displaystyle\oint_C f(z)dz$ if C is the circle $|z| = 3$ and $f(z)$ is:

   (a) $\dfrac{1}{z(z+1)}$

(b) $\dfrac{z+1}{z(z-1)}$

(c) $\dfrac{z}{(z+2)^2}$

(d) $\dfrac{1}{z(z+1)^3}$

(e) $\dfrac{z}{(z+1)(z+2)^2}$

6. Represent the function $\dfrac{z-1}{z+1}$ by

   (a) its MacLaurin series and state the region of convergence.

   (b) its Laurent series for the domain $|z| > 1$.

7. Obtain the expansion of the function $\dfrac{z+1}{z^2}$ into

   (a) its Taylor series about $z = 1$ and state the region of convergence;

   (b) its Laurent series for the domain $|z - 1| > 1$.

8. Obtain the first few non-zero terms of the Laurent series expansions of each of the functions $f(z)$:

   (a) $f(z) = \operatorname{cosec} z$

   (b) $f(z) = \dfrac{e^z}{z(z^2+1)}$

   (c) $f(z) = \dfrac{1}{e^z - 1}$

Chapter 4

The theory of residues

4.1 Introduction

We have seen in Chapter 2, that if a function $f(z)$ is analytic everywhere inside and on a simple closed contour C, then by the Cauchy-Goursat theorem, we have that $\int_C f(z)\, dz = 0$. However, if the function fails to be analytic at a finite number of points inside C, then we shall see in this chapter that the contribution to the value of the integral will come from the residues of the function at those points; the formal definition will be given in the next section. Before we define a residue, we define the following important quantities:

4.1.1 Isolated singular point

If a function $f(z)$ is analytic except at a point z_0, then the point z_0 is called an isolated singular point. As for example, $f(z) = 1/(z - z_0)$ is not analytic at $z = z_0$ and hence $z = z_0$ is an isolated singular point.

4.1.2 Poles

The principal part of the Laurent series expansion of finite numbers of series about an isolated singular point z_0 can be written as

$$f(z) = \frac{b_1}{z - z_0} + \frac{b_2}{(z - z_0)^2} + \cdots + \frac{b_m}{(z - z_0)^m}$$

when $0 < |z - z_0| < r$ for some positive number r, where $b_m \neq 0$. Then the isolated singular point z_0 is called a pole of order m of the function $f(z)$.

A pole of order $m = 1$ is called a simple pole.

4.1.3 Essential singular points

When the principal part of the Laurent series expansion of $f(z)$ about an isolated singular point z_0 has an infinite number of terms, then the point z_0 is called an **essential singular point.**

As for example,

$$f(z) = e^{1/z} \;=\; 1 + \sum_{n=1}^{\infty} \frac{1}{n! \, z^n}$$

is not analytic at $z = 0$ and hence $z = 0$ is an essential singular point. As $z \to 0$, each term (except the first) in the expansion $\to 0$, and one might be tempted to regard the singularity as a pole of infinite order. There is, however, a considerable difference between the behaviour of a function near a pole and that near such an essential singular point. This can be demonstrated by examining the behaviour of the function $w = e^{1/z}$ as $z \to 0$ as follows. We know that
$(1/z) = (1/r)(\cos\theta - i\sin\theta)$, so $w = e^{(\cos\theta/r)}\{\cos(\sin\theta/r) - i\sin(\sin\theta/r)\}$. Therefore in the w plane we have $w = u + iv = r_1 e^{-i\theta_1}$, where $r_1 = e^{(\cos\theta/r)}$, $\theta_1 = (\sin\theta/r)$, $u = r_1\cos\theta_1$, and $v = -r_1\sin\theta_1$.

Choosing a value θ_0 in $0 < \theta < (\pi/2)$, it can be easily observed that $r_1 = e^{(\cos\theta_0/r)}$ and $\theta_1 = (\sin\theta_0/r)$ increase rapidly in the w plane as $r \to 0$. Accordingly, the vector $r_1(\cos\theta_1 - i\sin\theta_1)$, corresponding to w, rotates with ever-increasing length and angular velocity, whereas near a pole, the angle in the w plane remains fixed. Thus the representative point of w traces out a spiral of ever increasing radius. Hence $e^{1/z}$ assumes an infinity of values as $z \to 0$, for any θ_0 in $0 < \theta < (\pi/2)$ in the z plane. Similar behaviour of this function can be observed for any values in $0 < \theta < 2\pi$ in the z plane. This is a remarkable difference between a pole and an essential singularity.

4.1.4   Removable singularity

A function that is not analytic at a point z_0, but that can be made analytic thereby merely assigning a suitable value to the function at that point, is said to have a removable singular point at z_0. In actual practice, $\lim_{z \to z_0} f(z)$ should exist for such a function. As for example, $z = 0$ is a removable singular point of $f(z) = (\sin z/z)$, because $\lim_{z \to 0} (\sin z/z) = 1$.

4.1.5   Branch points and branch cuts

Consider the transformation $w = z^{1/2}$ where $z = re^{i\theta}$. Therefore $w = r^{1/2}e^{i\theta/2}$. When z makes one complete circuit, w makes only half of the circuit (see Fig. 4.1). Thus

$$w = \sqrt{r}\, e^{1/2(\theta + 2\pi)i} = -\sqrt{r}e^{i\theta/2} \tag{4.1}$$

Again, when z makes two complete circuits,

$$w = -\sqrt{r}e^{i/2(\theta + 2\pi)} = \sqrt{r}\, e^{i\theta/2} \tag{4.2}$$

Hence, w has two distinct values corresponding to only one value of z. And w is called a multiple valued function.

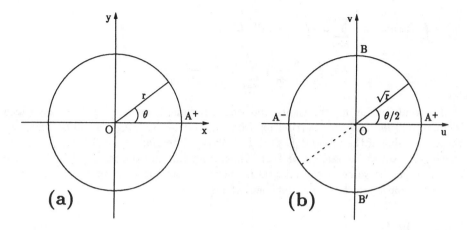

Fig. 4.1: Definition of branch points and branch cuts.

In this example, the origin O is defined as **a branch point**; and the line A^-A^+ in the w-plane is called **a branch cut**, which is a dividing line between the two branches A^+BA^- and $A^-B'A^+$. But in actual practice, a branch cut is in the z-plane along the x axis as shown in Fig. 4.1(a). It is to be noted that in each branch, the function is analytic. A branch point is usually regarded as a singular point.

4.2 Definition of the residue

If the function $f(z)$ is represented by the Laurent series in the domain $0 < |z - z_0| < r$, as follows:

$$f(z) = \sum_{n=0}^{\infty} a_n(z - z_0)^n + \frac{b_1}{z - z_0} + \frac{b_2}{(z - z_0)^2} + \cdots \qquad (4.3)$$

where

$$b_1 = \frac{1}{2\pi i} \int_C f(z)\, dz \qquad (4.4)$$

in which C is any closed contour around z_0 described in the positive sense such that $f(z)$ is analytic within and on C except at the point $z = z_0$ itself. Then, b_1, the coefficient of $1/(z - z_0)$ is defined as the residue of $f(z)$ at the isolated singular point z_0. The coefficient b_1 can be obtained by the formula of Laurent series coefficients as already demonstrated in the last chapter. Otherwise, we can determine b_1 by *term by term* integration of $f(z)$ with respect to z around the closed contour C by using the Cauchy's integral formula. If we do so, we see that

$$\int_C f(z)dz = \sum_{n=0}^{\infty} a_n \int_C (z - z_0)^n \, dz$$

$$+ b_1 \int_C \frac{dz}{z - z_0} + \sum_{n=2}^{\infty} b_n \int_C \frac{dz}{(z - z_0)^n}.$$

The contribution for the right-hand side will come from $b_1 \int_C dz/(z - z_0)$ which is equal to $2\pi i b_1$, and the contribution from the rest of the integrals will be zero, and hence the value of the integral $\int_C f(z) \, dz = 2\pi i b_1$. Therefore, $b_1 = (1/2\pi i) \int_C f(z) \, dz$ which is known as the residue of $f(z)$. Thus, if we can find the residue b_1 at the pole $z = z_0$, then the value of the integral is known. From the Laurent series expansion (3.15), we know that b_1 is the coefficient of $1/(z - z_0)$.

Example 1

Find the residue of $f(z) = (z + 1)/\{z^2(z - 2)\}$
    (a)   at $z = 0$
    (b)   at $z = 2$

Solution

(a) $f(z)$ is expanded about $z = 0$ by Taylor's expansion which yields

$$f(z) = \frac{z+1}{-2z^2}(1 - z/2)^{-1}$$

$$= \frac{z+1}{-2z^2}\left(1 + \frac{z}{2} + \frac{z^2}{4} + \frac{z^3}{8} + \cdots\right), \qquad 0 < |z| < 2$$

This can be rewritten as

$$f(z) = -(1 + z)\left(\frac{1}{2z^2} + \frac{1}{4z} + \frac{1}{8} + \frac{z}{16} + \cdots\right)$$

$$= -\left[\frac{1}{2z^2} + \frac{1}{4z} + \frac{1}{8} + \frac{z}{16} + \cdots + \frac{1}{2z} + \frac{1}{4} + \frac{z}{8} + \frac{z^2}{16} + \cdots\right]$$

$$= -\frac{1}{2z^2} - \frac{3}{4z} - \frac{3}{8} - \frac{3z}{16} - \cdots.$$

Thus, the residue of $f(z)$ at $z = 0$, i.e. the coefficient of $1/z$ in the expansion of $f(z)$ about $z = 0$ in the last expression is $-\frac{3}{4}$.
    (b)   If we expand $f(z)$ about $z = 2$, we obtain:

$$f(z) = \frac{z+1}{z-2}\frac{1}{(z-2+2)^2}$$

$$= \frac{z+1}{z-2} \frac{1}{4} \left(1 + \frac{z-2}{2}\right)^{-2}, \qquad \text{for } |z-2| < 2$$

$$= \frac{(z+1)}{4(z-2)} \left[1 - 2\frac{(z-2)}{2} + 3\left(\frac{z-2}{2}\right)^2 - 4\left(\frac{z-2}{2}\right)^3 + \cdots\right]$$

$$= (1+z) \left[\frac{1}{4(z-2)} - \frac{1}{4} + \frac{3}{16}(z-2)\cdots\right]$$

$$= [(z-2)+3][\frac{1}{4(z-2)} - \frac{1}{4} + \frac{3}{16}(z-2)\cdots]$$

$$= \frac{3}{4(z-2)} - \frac{3}{4} + \frac{9}{16}(z-2) + \cdots + \frac{1}{4} - \frac{1}{4}(z-2) + \cdots$$

Thus, the residue of $f(z)$ at $z = 2$, i.e. the coefficient of the term $1/(z-2)$ in the last expression is $\frac{3}{4}$.

Example 2

Find the residue of $f(z) = e^{1/z}$ about $z = 0$.

Solution

In this case, $z = 0$ is an essential sigular point. Expanding $f(z)$ about $z = 0$ by Laurent series expansion:

$$\begin{aligned} f(z) &= e^{1/z} \\ &= 1 + \frac{1}{z} + \frac{1}{2!z^2} + \frac{1}{3!z^3} + \cdots \qquad |z| > 0 \end{aligned}$$

Here the coefficient of $1/z$ is unity and hence the residue of $f(z)$ at $z = 0$ is unity.

4.3   Residue theorem

Theorem 4.1

If C is a simple closed contour and if a function $f(z)$ is analytic within and on C except for a finite number of singular points $z_1, z_2, \cdots z_n$ interior to C then,

$$\int_C f(z)\, dz = 2\pi i[R_1 + R_2 + \cdots + R_n]$$

where $R_1, R_2, \cdots R_n$ are the residues of $f(z)$ at its singular points within C.

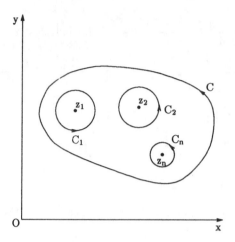

Fig. 4.2: Circle $C_1, C_2, \cdots C_n$ enclosing the singular points $z_1, z_2, \cdots z_n$ within a simple closed contour C.

Proof

Suppose the points z_k be the centres of positively oriented circles C_k which are interior to C and so small that no two of the circles have points in common (see Fig. 4.2). These circles, together with the simple closed contour C form the boundary of a domain called the multiply connected domain throughout which f is analytic.

Hence, by the extension of the Cauchy-Goursat theorem, we obtain

$$\int_C f(z)\, dz \; - \; \int_{C_1} f(z)\, dz - \int_{C_2} f(z)\, dz - \cdots - \int_{C_n} f(z)\, dz = 0$$

or, $$\int_C f(z)\, dz \;=\; \int_{C_1} f(z)\, dz + \int_{C_2} f(z)\, dz + \cdots + \int_{C_n} f(z)\, dz \qquad (4.5)$$

Now we know from (4.4) that:

$$2\pi i R_k = \int_{C_k} f(z)\, dz \qquad k = 1, 2, \cdots n$$

and hence,

$$\int_C f(z)\, dz = 2\pi i [R_1 + R_2 + \cdots + R_n] \qquad (4.6)$$

Hence this is the required proof.

To find the residues by the use of series expansions may be tedious and sometimes very difficult, as we have seen in the last section. Therefore, we state the following theorem to find the residues very quickly and easily.

Theorem 4.2

If $f(z)$ has a pole of order $(m+1)$ at $z = z_0$, then the residue of $f(z)$ at $z = z_0$, is:

$$R_1 = b_1 = \lim_{z \to z_0} \left[\frac{1}{m!} \frac{d^m}{dz^m} \{(z - z_0)^{m+1} f(z)\} \right]$$

Proof

To prove the theorem, we assume first that $f(z)$ has a simple pole or a first-order pole at $z = z_0$. It follows that we can expand $f(z)$ in the following way:

$$f(z) = \frac{b_1}{z - z_0} + a_0 + a_1(z - z_0) + a_2(z - z_0)^2 + \cdots \tag{4.7}$$

$$b_1 = \text{residue at the pole } z = z_0$$

$$= \text{coefficient of } \frac{1}{z - z_0} \tag{4.8}$$

Thus, if we multiply (4.7) by $(z - z_0)$

$$(z - z_0)f(z) = b_1 + a_0(z - z_0) + a_1(z - z_0)^2 + \cdots$$

If we let $z \to z_0$, then

$$b_1 = \lim_{z \to z_0} \{(z - z_0)f(z)\} \tag{4.9}$$

Now, if $f(z)$ has a second-order pole at $z = z_0$, then $f(z)$ can be expressed:

$$f(z) = \frac{b_2}{(z - z_0)^2} + \frac{b_1}{(z - z_0)} + a_0 + a_1(z - z_0) + \cdots \tag{4.10}$$

To obtain the residue b_1, multiply (4.9) by $(z - z_0)^2$, and we get

$$(z - z_0)^2 f(z) = b_2 + b_1(z - z_0) + a_0(z - z_0)^2 + a_1(z - z_0)^3 + \cdots$$

and then differentiate with respect to z and then let $z \to z_0$

$$b_1 = \lim_{z \to z_0} \frac{d}{dz}[(z - z_0)^2 f(z)] \tag{4.11}$$

Similarly, if $f(z)$ has a multiple pole of order $(m+1)$, then $f(z)$ can be expressed as:

$$f(z) = \frac{b_{m+1}}{(z - z_0)^{m+1}} + \frac{b_m}{(z - z_0)^m} + \cdots + \frac{b_1}{z - z_0} + a_0 + a_1(z - z_0) + \cdots \tag{4.12}$$

To obtain residue b_1, multiply (4.11) by $(z - z_0)^{m+1}$, getting

$$(z - z_0)^{m+1} f(z) = b_{m+1} + b_m(z - z_0) + b_1(z - z_0)^m + a_0(z - z_0)^{m+1} + \cdots$$

and then differentiate m times with respect to z and then let $z \to z_0$:

$$b_1 = \lim_{z \to z_0} \frac{1}{m!} \frac{d^m}{dz^m} [(z - z_0)^{m+1} f(z)] \tag{4.13}$$

Hence the theorem is proved.

Another method for finding the residues of a function $f(z)$ at a pole z_0 is available if f can be written as a quotient.

$$f(z) = \frac{p(z)}{q(z)} \tag{4.14}$$

where p and q are both analytic at z_0 and $p(z_0) \neq 0$.

If the function $f(z)$ has a simple pole at z_0, in addition to the condition given there, $q(z_0) = 0$ and $q'(z_0) \neq 0$, then the residue of f at the simple pole z_0 is given by the formula

$$R_1 = \frac{p(z_0)}{q'(z_0)} \tag{4.15}$$

To prove this statement, we expand each of the analytic functions p and q into a Taylor series valid in a disk $|z - z_0| < r$ and we write:

$$f(z) = \frac{p(z_0) + (z - z_0)p'(z_0) + (1/2!)(z - z_0)^2 p''(z_0) + \cdots}{q(z_0) + (z - z_0)q'(z_0) + (1/2!)(z - z_0)^2 q''(z_0) + \cdots}$$

If $q(z_0) = 0$, then:

$$(z - z_0)f(z) = \frac{p(z_0) + (z - z_0)p'(z_0) + \cdots}{q'(z_0) + \{(z - z_0)/2!\}q''(z_0) + \cdots} \tag{4.16}$$

$$\lim_{z \to z_0} \{(z - z_0)f(z)\} = \frac{p(z_0)}{q'(z_0)}$$

which is the residue of $f(z)$ at the simple pole $z = z_0$.

Suppose that $q'(z_0) = 0$, then from (4.15) we have:

$$\frac{1}{1!}(z - z_0)^2 f(z) = \frac{p(z_0) + (z - z_0)p'(z_0) + \cdots}{q''(z_0)/2 + \{(z - z_0)/6\}q'''(z_0) + \cdots}$$

Now the residue of $f(z)$ at the double pole $z = z_0$ can be obtained as

$$
\begin{aligned}
R_1 &= \text{residue} \\
&= \lim_{z \to z_0} \frac{1}{1!} \frac{d}{dz} [(z - z_0)^2 f(z)] \\
&= \lim_{z \to z_0} \frac{d}{dz} \left[\frac{p(z_0) + (z - z_0)p'(z_0) + \cdots}{\frac{q''(z_0)}{2} + \frac{(z-z_0)}{6}q'''(z_0) + \cdots} \right] \\
&= \lim_{z \to z_0} \frac{(\frac{q''}{2} + \frac{z-z_0}{6}q''' + \cdot)(p'(z_0) + \cdots) - [(p(z_0) + (z - z_0)p'(z_0))\frac{q'''(z_0)}{6}]}{[\frac{q''(z_0)}{2} + \frac{(z-z_0)}{6}q'''(z_0)]^2} \\
&= \frac{2p'(z_0)}{q''(z_0)} - \frac{2p(z_0)q'''(z_0)}{3[q''(z_0)]^2} \tag{4.17}
\end{aligned}
$$

This formula is valid when the poles are of order $m = 2$. When $m > 2$, the corresponding formula for the residues are lengthy.

Example 1

If C is the circle $|z| = 4$, evaluate $\int_C f(z)\, dz$ for each of the following functions:

(a) $\dfrac{z}{z^2 - 1}$ (b) $\dfrac{1}{z(z-2)^3}$ (c) $\dfrac{z^2}{(z^2 + 3z + 2)^2}$

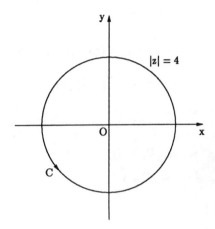

Fig. 4.3: A closed contour.

Solution

The contour C is depicted in Fig. 4.3.

(a) $\displaystyle \oint_C \frac{z}{z^2 - 1}\, dz$

The poles are at $z = \pm 1$ which are inside the given contour C. Thus we need to calculate two residues at these two poles:

$$R_1 = \text{residue at the pole } z = 1$$

$$= \lim_{z \to 1} \frac{z}{(z + 1)} = \frac{1}{2}$$

$$R_2 = \text{residue at the pole } z = -1$$

$$= \lim_{z \to -1} \frac{z}{(z - 1)} = \frac{1}{2}$$

$$\oint_C \frac{z\, dz}{z^2 - 1} = 2\pi i(R_1 + R_2) = 2\pi i \left(\frac{1}{2} + \frac{1}{2} \right) = 2\pi i$$

(b) $\qquad \oint_C \dfrac{dz}{z(z-2)^3}$

In this case, $z = 0$ is a simple pole, $z = 2$ is a triple pole. These poles are inside the contour C. Thus:

$$
\begin{aligned}
R_1 &= \text{residue at the pole } z = 0 \\
&= \lim_{z \to 0} \frac{z}{z(z-2)^3} = -\frac{1}{8}
\end{aligned}
$$

$$
\begin{aligned}
R_2 &= \text{residue at the triple pole } z = 2 \\
&= \lim_{z \to 2} \frac{1}{2!} \frac{d^2}{dz^2} \left\{ \frac{(z-2)^3}{z(z-2)^3} \right\} \\
&= \lim_{z \to 2} \frac{1}{2!} \frac{d^2}{dz^2} \left(\frac{1}{z} \right) = \frac{1}{2} \lim_{z \to 2} \frac{2}{z^3} = \frac{1}{8}
\end{aligned}
$$

$$
\begin{aligned}
\oint_C \frac{dz}{z(z-2)^3} &= 2\pi i (R_1 + R_2) \\
&= 2\pi i(0) = 0
\end{aligned}
$$

(c)

$$
\begin{aligned}
&\oint_C \frac{z^2}{(z^2 + 3z + 2)^2} dz \\
= &\oint_C \frac{z^2 dz}{(z+2)^2 (z+1)^2}
\end{aligned}
$$

Now $z = -1$ is a double pole and $z = -2$ is also a double pole inside the contour C. Therefore,

$$
\begin{aligned}
R_1 &= \text{residue at the double pole } z = -1 \\
&= \lim_{z \to -1} \frac{1}{1!} \frac{d}{dz} \left\{ \frac{z^2}{(z+2)^2} \right\} \\
&= \lim_{z \to -1} \left[\frac{2z}{(z+2)^2} - \frac{2z^2}{(z+2)^3} \right] = -4
\end{aligned}
$$

$$
\begin{aligned}
R_2 &= \text{residue at the double pole } z = -2 \\
&= \lim_{z \to -2} \frac{1}{1!} \frac{d}{dz} \left\{ \frac{z^2}{(z+1)^2} \right\} \\
&= \lim_{z \to -2} \left[\frac{2z}{(z+1)^2} - \frac{2z^2}{(z+1)^3} \right] = 4
\end{aligned}
$$

$$\oint_C \frac{z^2 dz}{(z^2 + 3z + 2)^2} = 2\pi(R_1 + R_2)$$
$$= 2\pi i \times 0$$
$$= 0$$

Example 2

If C is the circle $|z| = 2$, evaluate the following integrals:

(a) $\oint_C \tan z \, dz$ (b) $\oint_C \frac{dz}{\sinh 2z}$ (c) $\oint_C \frac{\cosh \pi z \, dz}{z(z^2 + 1)}$

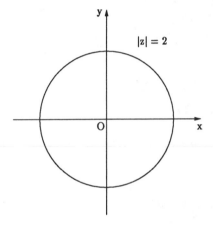

Fig. 4.4: A closed contour.

Solution

(a)

$$\oint_C \tan z \, dz$$
$$= \oint_C \frac{\sin z}{\cos z} dz$$

The contour C is shown in Fig. 4.4. $z = \pm\frac{\pi}{2}$ are the two poles inside the contour C.

Then,

$$R_1 = \text{residue at } z = \frac{\pi}{2}$$

$$= \lim_{z \to \pi/2} \left(z - \frac{\pi}{2} \right) \frac{\sin z}{\cos z}$$

$$= \lim_{z \to \pi/2} \frac{\sin z}{(\cos z)/(z - \frac{\pi}{2})}$$

$$= \lim_{z \to \pi/2} \frac{\sin z}{-\sin z} = -1$$

$$R_2 = \text{residue at } z = -\frac{\pi}{2}$$

$$= \lim_{z \to \pi/2} \left(z + \frac{\pi}{2} \right) \frac{\sin z}{\cos z}$$

$$= \lim_{z \to -\pi/2} \frac{\sin z}{-\sin z} = -1$$

Therefore

$$\oint_C \tan z \, dz = 2\pi(R_1 + R_2)$$

$$= -4\pi i$$

(b) In this case, $z = ik\pi/2$, where $k = 0, \pm 1$ are three simple poles which are inside the contour.

Then,

$$R_1 = \text{residue at } z = 0$$

$$= \lim_{z \to 0} \left(\frac{z}{\sinh 2z} \right) = \lim_{z \to 0} \frac{1}{(\sinh 2z)/z}$$

$$= \lim_{z \to 0} \left[\frac{1}{2 \cosh 2z} \right] = \frac{1}{2}$$

Similarly, we get $R_2 = $ residue at $z = i\pi/2$ is equal to $-1/2$ and $R_3 = $ residue at $z = -i\pi/2$ is equal to $-1/2$. Therefore,

$$\oint_C \frac{dz}{\sinh 2z} = (2\pi i) \left(\frac{1}{2} - \frac{1}{2} - \frac{1}{2} \right) = -\pi i$$

(c) In this case, $z = 0$ and $\pm i$ are all simple poles.

Then,

$$R_1 = \text{residue at } z = 0$$

$$= \lim_{z \to 0} \frac{\cosh \pi z}{(z^2 + 1)} = 1$$

$$R_2 = \text{residue at } z = i$$
$$= \lim_{z \to 0} \frac{\cosh \pi z}{z(z+i)} = \frac{\cosh \pi i}{(i)(2i)} = \frac{\cos \pi}{-2} = \frac{1}{2}$$

$$R_3 = \text{residue at } z = -i$$
$$= \lim_{z \to -i} \frac{\cosh \pi z}{z(z-i)} = \frac{\cosh \pi i}{-2} = \frac{\cos \pi}{-2} = \frac{1}{2}$$

Therefore,

$$\oint_C \frac{\cosh \pi z}{z(z^2+1)} dz$$
$$= 2\pi i (R_1 + R_2 + R_3)$$
$$= 4\pi i$$

Exercises

1. Expand each of the following functions at an isolated singular point by a Laurent series expansion. Determine if that point is a pole, an essential singular point, or a removable singular point of the given function.

(a) $\dfrac{z}{1+z^2}$

(b) $\dfrac{z^2}{1+z}$

(c) $z e^{1/z}$

(d) $\dfrac{\sin z}{z}$

(e) $\dfrac{\cos z}{z}$

(f) $z e^{1/z^2}$

2. Determine the order m of each pole of the following functions and then find the corresponding residues.

(a) $\dfrac{e^z}{z^2+1}$

(b) $\cot z$

(c) $\dfrac{1+z}{z^2-2z+1}$

(d) $\dfrac{z+1}{z^2(z-2)}$

(e) $\dfrac{z}{\cos z}$

(f) $\dfrac{1}{z - \sin z}$

3. Find the residues of $f(z)$ at $z = 0$:

(a) $f(z) = \operatorname{cosec}^2 z$

(b) $f(z) = \dfrac{\operatorname{cosec}(z^2)}{z^3}$

(c) $f(z) = z \cos\left(\dfrac{1}{z}\right)$

4. Find the residue of $f(z) = \dfrac{z}{z^2 + 4z + 8}$ at each of its poles.

5. Find the residue of $f(z) = \dfrac{3z^3 + 2}{(z - 1)(z^2 + 9)}$ at each of its poles.

6. What is the residue of $f(z) = \dfrac{1}{(z + 1)^3}$ at $z = -1$?

7. What is the residue of $f(z) = \dfrac{z}{1 - e^z}$ at $z = 0$?

8. If C is the circle $|z| = 3$, evaluate $\displaystyle\oint_C f(z)\,dz$ for each of the following functions:

(a) $\dfrac{3z^2 + 1}{(z - 1)(z^2 + 4)}$

(b) $\dfrac{z + 1}{z^2(z + 2)}$

(c) $\dfrac{z + 2}{z^3(z + 1)}$

(d) $\dfrac{1}{z(z - 2)^3}$

(e) $\dfrac{z}{(z^2 + 3z + 2)^2}$

(f) $\dfrac{1}{z^2 + z + 1}$

(g) $\dfrac{z + 1}{z(z^2 + 2z + 5)}$

9. If C is the unit circle, evaluate $\displaystyle\oint_C f(z)\,dz$ for each of the following functions:

(a) $z^{-1}e^z$

(b) $z^{-2}e^{-z}$

(c) $z^{-2}\sec z$

(d) $z\,e^{1/z^2}$

10. If C is the circle $|z| = 2$, evaluate $\oint_C f(z)dz$ for each of the following functions by finding the residues using the Laurent series expansion about the appropriate poles:

(a) $\dfrac{5z - 2}{z(z - 1)}$

(b) $z\,e^{1/z^2}$

(c) $\dfrac{z + 1}{z^2(z - 1)}$

4.4 Evaluation of real integrals

The evaluation of certain types of real definite integrals can be made by the theory of residues. This theory will be illustrated first in connection with the evaluation of the following types of integrals, namely:

- Evaluation of improper real integrals.

- Improper integrals involving trigonometric functions.

- Definite integrals of trigonometric functions.

- Integration around a branch point.

4.4.1 Evaluation of improper integrals

Consider a real improper integral

$$\int_{-\infty}^{\infty} f(x)dx \tag{4.18}$$

where $f(x)$ is continuous for all x. This integral may be written as:

$$\int_{-\infty}^{\infty} f(x)dx = \lim_{R_1 \to \infty} \int_{-R_1}^{0} f(x)dx + \lim_{R_2 \to \infty} \int_{0}^{R_2} f(x)dx \tag{4.19}$$

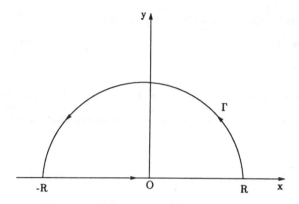

Fig. 4.5: A semi-circular contour C.

This is said to converge when the individual limits exist. If both limits exist, we may combine the two independent passages to $-\infty$ and ∞, and write:

$$\text{P.V.} \int_{-\infty}^{\infty} f(x)\ dx = \lim_{R \to \infty} \int_{-R}^{R} f(x)dx \tag{4.20}$$

provided the limit on the right exists. Here P.V. stands for the **Cauchy principal value** and is defined by the equation (4.20). It is to be noted here that the Cauchy principal value of the integral may exist even if the limits in (4.19) do not exist. For example:

$$\lim_{R \to \infty} \int_{-R}^{R} x dx = \lim_{R \to \infty} \left(\frac{R^2}{2} - \frac{R^2}{2} \right) = 0 \tag{4.21}$$

But,

$$\lim_{R \to \infty} \int_{0}^{R} x dx = \infty$$

and $$\quad \lim_{R \to \infty} \int_{-R}^{0} x dx = \infty \tag{4.22}$$

However, if the limits in (4.19) exist, then it is guaranteed that the Cauchy principal value in (4.20) does exist.

Now suppose the function $f(x)$ in (4.19) is a rational function and can be written as $f(x) = p(x)/q(x)$ where $p(x)$ and $q(x)$ are real polynomials with no factors in common and $q(x)$ has no real zero. If the degree of $q(x)$ is at least two greater than the degree of $p(x)$, the integral converges and the value of the integral can be found quite easily by determining its **Cauchy principal value.**

We now consider the corresponding contour integral:

$$\int_C f(z)dz \tag{4.23}$$

and around a path C as shown in Fig. 4.5.

Since $f(x)$ is rational, $f(z)$ has infinitely many poles in the upper half plane, and if we choose R large enough so that C encloses all these poles. By the residue theorem we then obtain:

$$\int_C f(z)dz = \int_\Gamma f(z)dz + \int_{-R}^R f(x)dx$$
$$= 2\pi i \sum \text{ residues of } f(z) \text{ at all poles within } C. \tag{4.24}$$

From this equation we have:

$$\int_{-R}^R f(x)dx = 2\pi i \sum \text{residues} - \int_\Gamma f(z)dz \tag{4.25}$$

We next prove that if $R \to \infty$, the value of the integral over the semi-circle Γ approaches zero. To do this, put $z = Re^{i\theta}$ and $dz = Rie^{i\theta}d\theta$ where θ varies from 0 to π in the semi-circle Γ. Since the degree of denominator of $f(z)$ is at least two units higher than the degree of numerator, we have:

$$|f(z)| < \frac{M}{|z|^2}$$

where $|z| = R$, and M is a constant.

Then

$$\left| \int_\Gamma f(z)dz \right| < \int_\Gamma |f(z)||dz|$$
$$< \frac{M}{R^2}\pi R$$
$$= \frac{M\pi}{R} \tag{4.26}$$

Hence, as $R \to \infty$ the value of the integral over Γ approaches zero. Thus (4.24) yields:

$$\int_{-\infty}^\infty f(x)dx = 2\pi i \sum \text{Residues}.$$

We shall demonstrate in the following examples the application of this theory.

Example 1

Establish the following integration formula with the aid of residues:

$$\int_0^\infty \frac{dx}{(1+x^2)^2} = \frac{\pi}{4}$$

Solution

Let us rewrite:

$$\int_0^\infty \frac{dx}{(1+x^2)^2} = \frac{1}{2}\int_{-\infty}^\infty \frac{dx}{(1+x^2)^2}$$

$$= \frac{1}{2}\int_{-\infty}^\infty f(x)\,dx$$

where

$$f(x) = \frac{1}{(1+x^2)^2}$$

Now consider the corresponding function $f(z) = 1/(1+z^2)^2$. The contour is the semi-circle as shown in Fig. 4.5. Here, $z = \pm i$ are the double poles of this function. Only pole $z = i$ is inside the contour.

Therefore, the residue R_1 at $z = i$ (a double pole) is given by

$$R_1 = \lim_{z \to i} \frac{d}{dz}\left\{\frac{1}{(z+i)^2}\right\} = \lim_{z \to i}\frac{-2}{(z+i)^3} = \frac{1}{4i}$$

Hence,

$$\int_{-\infty}^\infty f(x)\,dx = 2\pi i R_1 = \frac{\pi}{2}$$

Thus,

$$\int_0^\infty f(x)\,dx = \frac{1}{2}\int_{-\infty}^\infty f(x)\,dx = \frac{\pi}{4}$$

Example 2

Establish the following integration formula with the aid of residues:

$$\int_0^\infty \frac{x^2\,dx}{(x^2+1)(x^2+4)} = \frac{\pi}{6}$$

Solution

Let us rewrite

$$\int_0^\infty \frac{x^2 dx}{(x^2+1)(x^2+4)} = \frac{1}{2}\int_{-\infty}^\infty \frac{x^2 dx}{(x^2+1)(x^2+4)}$$
$$= \frac{1}{2}\int_{-\infty}^\infty f(x)dx$$

Now consider the corresponding function:

$$f(z) = \frac{z^2}{(z^2+1)(z^2+4)}$$

The contour is the semi-circle as shown in Fig. 4.5. Here $z = \pm i, \pm 2i$ are the poles of the function $f(z)$. The poles $z = i, 2i$ are inside the contour as shown in Fig. 4.5. Thus, by Cauchy's residue theory,

$$\int_{-\infty}^\infty f(x)dx = 2\pi i(R_1 + R_2)$$

where,

$$R_1 = \text{residue at } z = i$$
$$= \lim_{z\to i} \frac{z^2}{(z+i)(z^2+4)} = \frac{i^2}{2i(i^2+4)} = -\frac{1}{6i}$$

$$R_2 = \text{residue at } z = 2i$$
$$= \lim_{z\to 2i} \frac{z^2}{(z^2+1)(z+2i)} = \frac{(2i)^2}{(1+(2i)^2)4i} = \frac{1}{3i}$$

Hence

$$\int_{-\infty}^\infty f(x)dx = (2\pi i)(\frac{-1}{6i} + \frac{1}{3i}) = (2\pi)(\frac{1}{6}) = \frac{\pi}{3}$$
$$\int_0^\infty f(x)dx = \frac{1}{2}\int_{-\infty}^\infty f(x)dx = \frac{\pi}{6}$$

4.4.2 Improper integrals involving trigonometric functions

The improper integrals of the form

$$\int_{-\infty}^\infty \frac{p(x)}{q(x)}\cos x dx \quad \text{and} \quad \int_{-\infty}^\infty \frac{p(x)}{q(x)}\sin x dx \tag{4.27}$$

where $p(x)$ and $q(x)$ are real polynomials and $q(x)$ has no zeros, can be evaluated by residue theory. However, the method of the previous section can not be used directly because of the fact that $|\cos z|$ and $|\sin z|$ increase like e^y as y tends to ∞. Therefore, the previous method must be modified as follows:

Noting that the integrals (4.27) are the real and imaginary parts of

$$\int_{-\infty}^{\infty} \frac{p(x)}{q(x)} e^{iz}\, dx \qquad\qquad (4.28)$$

and that $e^{iz} = e^{ix} \cdot e^{-y}$ when y goes to infinity $e^{iz} \to 0$, in the upper half plane; therefore (4.28) is the proper representation of (4.27). In the following we shall demonstrate this theory by examples.

Example 3

Evaluate the integral $\int_0^\infty \{\cos ax/(x^2+1)\}dx, \quad a \geq 0$.

Solution

We can rewrite

$$\int_0^\infty \frac{\cos ax}{x^2+1}dx \;=\; \frac{1}{2}Re\int_{-\infty}^{\infty}\frac{e^{iax}}{x^2+1}dx = \frac{1}{2}Re\int_{-\infty}^{\infty} f(x)dx$$

where $f(x) = e^{iax}/(x^2+1)$. The corresponding function is then $f(z) = e^{iaz}/(z^2+1)$.

The contour is the semi-circle as shown in Fig. 4.5.

Here $z = \pm i$ are the simple poles of this function, and $z = i$ is the only pole inside the contour. Therefore,

$$\int_{-\infty}^{\infty} f(x)dx = 2\pi i R_1$$

where the residue R_1 at the pole $z = i$ is given by

$$R_1 \;=\; \lim_{z\to i}\frac{e^{iaz}}{(z+i)} = \frac{e^{-a}}{2i}$$

Therefore we obtain

$$\int_{-\infty}^{\infty} f(x)dx \;=\; (2\pi i)\frac{e^{-a}}{2i} = \pi e^{-a}$$

Thus,

$$\int_0^\infty \frac{\cos ax}{x^2+1}dx \;=\; \frac{1}{2}Re\int_{-\infty}^{\infty} f(x)dx = \frac{\pi e^{-a}}{2}-$$

Example 4

Evaluate the integral

$$\int_{-\infty}^{\infty} \frac{\sin x \, dx}{x^2 + 4x + 5}$$

Solution

We can rewrite

$$\int_{-\infty}^{\infty} \frac{\sin x}{x^2 + 4x + 5} dx = Im \int_{-\infty}^{\infty} \frac{e^{ix} dx}{x^2 + 4x + 5}$$

$$= Im \int_{-\infty}^{\infty} f(x) dx$$

where $f(x) = e^{ix}/(x^2 + 4x + 5)$. The corresponding function is then $f(z) = e^{iz}/(z^2 + 4z + 5) = e^{iz}/\{(z + 2)^2 + 1\}$. Here $z = -2 \pm i$ are the simple poles of this function, and $z = -2 + i$ is the only pole inside the contour (see Fig. 4.5).

Therefore $\int_{-\infty}^{\infty} f(x) dx = 2\pi i R_1$ where R_1 is the residue at the pole at $z = -2 + i$. Thus

$$R_1 = \lim_{z \to -2+i} \frac{e^{iz}}{2z + 4} = \frac{e^{i(-2+i)}}{2(-2 + i) + 4} = \frac{e^{-2i}e^{-1}}{2i}$$

Hence we have

$$\int_{-\infty}^{\infty} f(x) dx = (2\pi i)\frac{e^{-1}e^{-2i}}{2i} = \pi e^{-1}(\cos 2 - i \sin 2)$$

Subsequently the value of the integral is

$$\int_{-\infty}^{\infty} \frac{\sin x}{x^2 + 4x + 5} dx = -\frac{\pi}{e} \sin 2$$

4.4.3   Definite integrals of trigonometric functions

In evaluating certain definite integrals of the type:

$$\int_{0}^{2\pi} g(\cos \theta, \sin \theta) d\theta \tag{4.29}$$

the method of residues is useful. Here $g(\cos \theta, \sin \theta)$ is a real rational function of $\cos \theta$ and $\sin \theta$ where θ varies from 0 to 2π.
Setting $e^{i\theta} = z$, we obtain

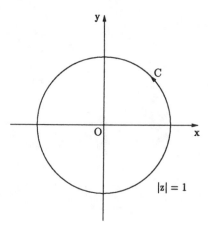

Fig. 4.6: A circle.

$$\cos\theta = \frac{1}{2}(e^{i\theta} + e^{-i\theta}) = \frac{1}{2}\left(z + \frac{1}{z}\right)$$

$$\sin\theta = \frac{1}{2i}(e^{i\theta} - e^{-i\theta}) = \frac{1}{2i}\left(z - \frac{1}{z}\right)$$

and we see that the integral becomes a rational function of z, say, $f(z)$. As θ varies from 0 to 2π, z traverses a unit circle $|z| = 1$ in the anti-clockwise sense (see Fig. 4.6).
    Thus

$$z = e^{i\theta}$$
$$dz = i\ e^{i\theta}d\theta$$
$$\quad = iz\ d\theta$$
$$\frac{dz}{iz} = d\theta$$

Substituting these values into the following contour integral yields,

$$\oint_C g\left(\frac{z + z^{-1}}{2}, \frac{z - z^{-1}}{2i}\right)\frac{dz}{iz}$$

with the integration being taken in the counter-clockwise sense around the unit circle. In the following, we shall give some illustrations of the use of this integral.

Example 5

Evaluate the following integral involving cosine and sine:

$$\int_0^{2\pi} \frac{d\theta}{2 + \cos\theta}$$

Solution

We know that

$$\cos\theta = \frac{e^{i\theta} + e^{-i\theta}}{2}$$

$$= \frac{1}{2}(z + z^{-1})$$

where $z = e^{i\theta}$ and $(dz/iz) = d\theta$.

Therefore,

$$\int_0^{2\pi} \frac{d\theta}{2 + \cos\theta}$$

$$= \oint_C \frac{1}{\{2 + \frac{1}{2}(z + \frac{1}{z})\}} \frac{dz}{iz}$$

$$= \oint_C \frac{2dz}{(4z + z^2 + 1)i}$$

$$= \frac{2}{i} \int_C \frac{dz}{(z + 2)^2 - (\sqrt{3})^2}$$

The poles of the function are at $z = -2 \pm \sqrt{3}$. Only pole $z = -2 + \sqrt{3}$ is inside the contour (see Fig. 4.6).

Thus

$$\int_0^{2\pi} \frac{d\theta}{2 + \cos\theta} = \frac{2}{i} \int_C \frac{dz}{z^2 + 4z + 1}$$

$$= (2\pi i)\left[\frac{2}{i} R_1\right] = 4\pi R_1$$

where

$$R_1 = \text{residue at the pole } z = -2 + \sqrt{3}$$

$$= \lim_{z \to -2+\sqrt{3}} \frac{1}{2z + 4}$$

$$= \frac{1}{2(-2 + \sqrt{3}) + 4} = \frac{1}{2\sqrt{3}}$$

$$\int_0^{2\pi} \frac{d\theta}{2 + \cos\theta} = (4\pi)\frac{1}{2\sqrt{3}}$$

$$= \frac{2\pi}{\sqrt{3}}$$

By elementary method of solution, we obtain

$$\int_0^{2\pi} \frac{d\theta}{2 + \cos\theta}$$

$$= 2\int_0^{\pi} \frac{d\theta}{2 + \cos\theta}$$

$$= 2\int_0^{\pi} \frac{d\theta}{2\cos^2\frac{\theta}{2} + 2\sin^2\frac{\theta}{2} + \cos^2\frac{\theta}{2} - \sin^2\frac{\theta}{2}}$$

$$= 2\int_0^{\pi} \frac{d\theta}{3\cos^2\frac{\theta}{2} + \sin^2\frac{\theta}{2}}$$

$$= 2\int_0^{\pi} \frac{\sec^2\frac{\theta}{2}d\theta}{3 + \tan^2\frac{\theta}{2}} \qquad \text{but} \quad \tan\theta/2 = z, \quad \sec^2\theta/2 \, d\theta = 2dz$$

$$= 2\int_0^{\infty} \frac{2dz}{3 + z^2}$$

$$= \frac{4}{\sqrt{3}}\tan^{-1}\frac{z}{\sqrt{3}}\Bigg]_0^{\infty}$$

$$= \left(\frac{4}{\sqrt{3}}\right)\left(\frac{\pi}{2}\right) = \frac{2\pi}{\sqrt{3}}$$

It is now evident that these two methods give identical solutions.

Example 6

Evaluate the following integral involving cosine and sine:

$$\int_0^{2\pi} \frac{\cos\theta}{3 + \sin\theta}d\theta$$

Solution

We know that

$$\cos\theta = \frac{1}{2}(e^{i\theta} + e^{-i\theta}) = \frac{1}{2}\left(z + \frac{1}{z}\right)$$

$$\sin\theta = \frac{1}{2i}(e^{i\theta} - e^{-i\theta}) = \frac{1}{2i}\left(z - \frac{1}{z}\right)$$

where $z = e^{i\theta}$ and $dz/iz = d\theta$.

Therefore,

$$\int_0^{2\pi} \frac{\cos\theta d\theta}{3 + \sin\theta}$$

$$= \int_C \frac{\frac{1}{2}(z + \frac{1}{z})}{3 + \frac{1}{2i}(z - \frac{1}{z})}\frac{dz}{iz}$$

$$= \int_C \frac{z^2 + 1}{6z + \frac{1}{i}(z^2 - 1)} \frac{dz}{iz}$$

$$= \int_C \frac{z^2 + 1}{(z^2 - 1 + 6iz)z} dz$$

$$= 2\pi i \sum \text{residues}$$

The poles are at $z = 0$ and at

$$z = \frac{-6i \pm \sqrt{-36 + 4}}{2}$$

$$= -3i \pm i2\sqrt{2} = z_1, z_2$$

The only poles inside the unit circle are at $z = 0$ and at $z = z_1 = (-3 + 2\sqrt{2})i$ (see Fig. 4.6).

$$R_1 = \text{residue at } z = 0$$
$$= \lim_{z \to 0} \left(\frac{z^2 + 1}{z^2 + 6iz - 1} \right) = -1$$
$$R_2 = \text{residue at } z = z_1$$
$$= \lim_{z \to z_1} \left(\frac{z^2 + 1}{z(z - z_2)} \right) = 1$$
$$\int_0^{2\pi} \frac{\cos\theta \; d\theta}{3 + \sin\theta} = 2\pi i(-1 + 1) = 0$$

In evaluating R_2, it is to be noted that $z_1 z_2 = -1$.

Hence we have,

$$\int_0^{2\pi} \frac{\cos\theta d\theta}{3 + \sin\theta} = 0$$

4.4.4   Integration around a branch point

The residue theorem can be used in evaluating real integrals involving branch points and branch cuts. We shall illustrate this theory considering an example in the following:

Example 7

Derive the formula:

$$\int_0^\infty \frac{\ln x}{(x^2 + 1)^2} dx = -\frac{\pi}{4}$$

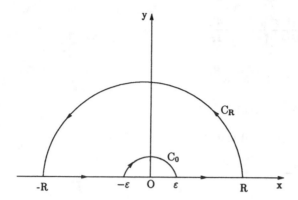

Fig. 4.7: A semi-circular contour with a branch point at $x = 0$.

Solution

We can rewrite

$$I = \int_0^\infty \frac{\ln x}{(x^2 + 1)^2} dx = \int_0^\infty f(x)\ dx$$

where $f(x) = \ln x / \{(x^2 + 1)^2\}$. The corresponding function is given by $f(z) = \ln z / \{(z^2 + 1)^2\}$.

To evaluate this integral, we consider the following contour where C_R and C_0 are the two simple closed contours as shown in Fig. 4.7.

Here $z = \pm i$ are the double poles and only $z = i$ is inside the closed contour. Also $z = 0$ is a branch point.

Now by theory of residues, we have,

$$\int_{C_R} f(z)\ dz + \int_{-R}^{-\epsilon} f(x)\ dx + \int_{C_0} f(z)\ dz + \int_{\epsilon}^{R} f(x)dx = 2\pi i R_1$$

where the residue R_1 at the pole $z = i$ is

$$R_1 = \lim_{z \to i} \frac{d}{dz} \frac{\ln z}{(z + i)^2}$$

$$= \lim_{z \to i} \frac{d}{dz} \left[\frac{\ln z}{(z + i)^2} \right]$$

$$= \lim_{z \to i} \left[\frac{1}{z(z + i)^2} - \frac{2 \ln z}{(z + i)^3} \right]$$

$$R_1 = \frac{1}{i(2i)^2} - \frac{2 \ln i}{(2i)^3} = \frac{1 - \ln i}{-4i}$$

Therefore,

$$\int_{C_R} f(z)dz + \int_{-R}^{-\epsilon} f(x)dx + \int_{C_0} f(z)dz + \int_{\epsilon}^{R} f(x)dx = \frac{(\ln i - 1)\pi}{2}$$

Now we know $z = Re^{i\theta}$ so that $dz = Rie^{i\theta}$ and $|z| = Rd\theta$. Therefore using this information

$$\left| \int_{C_R} f(z)dz \right|$$

$$\leq \int_{C_R} |f(z)||dz|$$

$$\leq \int_{C_R} \frac{|\ln z|}{|z^2 + 1|^2}|dz|$$

$$< \int_0^\pi \frac{\ln R + \theta}{(R^2 - 1)^2}R \ d\theta$$

$$= R\left[\frac{\pi \ln R}{(R^2 - 1)^2} + \frac{\pi^2}{2(R^2 - 1)^2} \right] = \frac{R\pi}{(R^2 - 1)^2}\left[\ln R + \frac{\pi}{2} \right]$$

When $R \to \infty$ and $\left| \int_{C_R} f(z)dz \right| \to 0$,

$$\int_{C_0} \frac{\ln z}{(z^2 + 1)^2}dz = \int_\pi^0 \frac{\ln(\epsilon e^{i\theta})i\epsilon e^{i\theta}d\theta}{(\epsilon^2 e^{2i\theta} + 1)^2}$$

$$= i\epsilon \int_\pi^0 \frac{\ln \epsilon + i\theta}{(\epsilon^2 e^{2i\theta} + 1)^2}e^{i\theta}d\theta$$

But we know $z = \epsilon e^{i\theta}$ such that $dz = \epsilon i e^{i\theta}d\theta$. As $\epsilon \to 0$ and $\int_{C_0} \frac{\ln z}{(z^2 + 1)^2}dz \to 0$,

we have

$$\int_{-R}^{-\epsilon} \frac{\ln x}{(x^2 + 1)^2}dx + \int_\epsilon^R \frac{\ln x}{(x^2 + 1)^2}dx = \frac{\pi}{2}(\ln i - 1)$$

The left-hand side can be written as:

$$\int_R^\epsilon \frac{\ln(-x)}{(x^2 + 1)^2}(-dx) + \int_\epsilon^R \frac{\ln x}{(x^2 + 1)^2}dx$$

$$= \int_\epsilon^R \frac{2\ln \ x + \pi i}{(x^2 + 1)^2}dx$$

$$= 2\int_\epsilon^R \frac{\ln x}{(x^2 + 1)^2}dx + \pi i \int_\epsilon^R \frac{dx}{(x^2 + 1)^2}$$

$$= 2\int_\epsilon^R \frac{\ln x}{(x^2 + 1)^2}dx + \pi i \left(\frac{\pi}{4} \right)$$

Therefore, when $\varepsilon \to 0$ and $R \to \infty$,

$$2\int_0^\infty \frac{\ln x}{(x^2+1)^2}dx \;+\; i\frac{\pi^2}{4} = \frac{\pi}{2}\left(\frac{\pi i}{2} - 1\right)$$

$$\int_0^\infty \frac{\ln x}{(x^2+1)^2}dx \;=\; -\frac{\pi}{4}$$

Example 8

Evaluate the following integral

$$\int_0^\infty \frac{x^k}{1+x^2}dx, \qquad -1 < k < 1$$

Solution

We rewrite the integrand as

$$f(x) \;=\; \frac{x^k}{1+x^2}$$

$$f(z) \;=\; \frac{z^k}{1+z^2}$$

A suitable contour is shown in Fig. 4.7. Thus by the theory of residue, we have,

$$\int_{C_R} f(z)dz + \int_{-R}^{-\varepsilon} f(x)dx + \int_{C_0} f(z)dz + \int_\varepsilon^R f(x)dx = 2\pi i \sum R_n$$

It can be easily shown that $\int_{C_R} f(z)dz \to 0$ as $R \to \infty$.
Using the relation $z = \varepsilon e^{i\theta}$, such that $dz = \varepsilon i e^{i\theta}d\theta$, we have

$$\int_{C_0} f(z)dz \;=\; \int_{C_0} \frac{z^k}{z^2+1}dz = \int_\pi^0 \frac{\varepsilon^k e^{ki\theta} \cdot \varepsilon i e^{i\theta}d\theta}{1+\varepsilon^2 e^{2i\theta}}$$

As $\varepsilon \to 0$, $\int_{C_0} f(z) \to 0$.

We know the poles of the function $f(z)$ are given by the roots of $z^2 + 1 = 0$. Thus the poles are at $z = \pm i$, only i is inside the closed contour. Hence, the residue at $z = i$ is given by

$$R_1 = \lim_{z \to i} \frac{z^k}{z+i} = \frac{i^k}{2i} = \frac{1}{2}i^{k-1} = \frac{1}{2}e^{(\pi/2)(k-1)i}$$

Thus,

$$\int_{-R}^{-\varepsilon} \frac{x^k}{1+x^2}dx + \int_\varepsilon^R \frac{x^k}{1+x^2}dx \;=\; \frac{2\pi i}{2i}e^{i(\pi/2)k} = \pi e^{i(\pi k/2)}$$

Hence when R tends to infinity and ϵ goes to zero, we have,

$$\int_0^\infty \frac{(-1)^k x^k}{1+x^2}dx + \int_0^\infty \frac{x^k}{1+x^2}dx = \pi e^{i(\pi k/2)}$$

Simplifying the above equation we have

$$\left(\int_0^\infty \frac{x^k}{1+x^2}dx\right)(1+e^{k\pi i}) = \pi e^{i\pi k/2}$$

Therefore,

$$\int_0^\infty \frac{x^k}{1+x^2}dx = \frac{\pi e^{k\pi i/2}}{1+e^{k\pi i}}$$

$$= \frac{\pi/2}{(e^{k\pi i/2} + e^{-k\pi i/2})/2}$$

$$= \frac{\pi}{2\cos(k\pi/2)}$$

Example 9

Derive the following integration formula:

$$\int_0^\infty \frac{x^k}{(1+x^2)^2}dx = \frac{\pi}{4}\frac{1-k}{\cos(k\pi/2)}$$

where $-1 < k < 1$.

Solution

We rewrite the integral:

$$\int_0^\infty \frac{x^k}{(x^2+1)^2}dx = \int_0^\infty f(x)dx$$

The corresponding $f(z)$ is given by:

$$f(z) = \frac{z^k}{(z^2+1)^2}$$

We choose the contour shown in Fig. 4.7 to evaluate this integral. Now, by the residue theory, we have,

$$\int_{C_R} f(z)dz + \int_{-R}^{-\epsilon} f(x)dx + \int_{C_O} f(z)dz$$

$$+ \int_\epsilon^R f(x)dx = 2\pi i\,(\text{residues})$$

$$= 2\pi i R_1$$

The only pole inside the contour is $z = i$, but $z = 0$ is the branch point of function. Thus,

$$
\begin{aligned}
R_1 &= \text{residue at the pole } z = i \,(\text{a double pole}) \\
&= \lim_{z \to i} \frac{d}{dz}\left(\frac{z^k}{(z+i)^2}\right) \\
&= \lim_{z \to i}\left[\frac{kz^{k-1}}{(z+i)^2} - \frac{2z^k}{(z+i)^3}\right] \\
&= \frac{1-k}{4i}e^{k\pi i/2}
\end{aligned}
$$

$$
\int_{C_R} f(z)dz + \int_{-R}^{-\epsilon} f(x)dx + \int_{C_0} f(z)dz + \int_{\epsilon}^{R} f(x)dx = \frac{\pi}{2}(1-k)e^{k\pi i/2}
$$

It can be easily shown that $|\int_{C_R} f(z)dz| \to 0$ as $R \to \infty$ and $|\int_{C_0} f(z)dz| \to 0$ as $\epsilon \to 0$ (see Example 7). Therefore,

$$
\int_{-R}^{-\epsilon} \frac{x^k}{(x^2+1)^2}dx + \int_{\epsilon}^{R} \frac{x^k}{(x^2+1)^2}dx = \frac{\pi}{2}(1-k)e^{k\pi i/2}
$$

or

$$
(1+e^{k\pi i})\int_0^{\infty} \frac{x^k}{(x^2+1)^2}dx = \frac{\pi}{2}(1-k)e^{k\pi i/2}
$$

$$
\int_0^{\infty} \frac{x^k}{(x^2+1)^2}dx = \frac{\pi}{2}(1-k)\frac{e^{k\pi i/2}}{1+e^{k\pi i}} = \frac{\pi}{4}\frac{1-k}{\cos(k\pi/2)}
$$

Example 10

Evaluate the integral $\int_0^{\infty}\{x^{-k}/(x+a)\}dx$, where a is a positive real number and $0 < k < 1$. This integral is important in the study of **the gamma function**.

Solution

We rewrite the integrand as

$$
f(x) = \frac{x^{-k}}{x+a}
$$

$$
f(z) = \frac{z^{-k}}{z+a}
$$

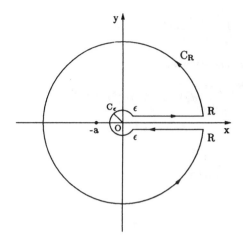

Fig. 4.8: A circular contour.

A suitable contour is shown in Fig. 4.8. Thus by the theory of residue, we have,

$$\int_\epsilon^R f(z)dz + \int_{C_R} f(z)dz + \int_R^\epsilon f(z)dz + \int_{C_\epsilon} f(z)dz = 2\pi i \sum R_n$$

It can be easily shown that $\int_{C_R} f(z)dz \to 0$ as $R \to \infty$.
Using the relation $z = \epsilon e^{i\theta}$, such that $dz = \epsilon i e^{i\theta} d\theta$, we have

$$\int_{C_\epsilon} f(z)dz = \int_{C_\epsilon} \frac{z^{-k}}{z+a} dz$$

$$= \int_{2\pi}^0 \frac{\epsilon^{-k} e^{-ki\theta} \cdot \epsilon i e^{i\theta} d\theta}{\epsilon e^{i\theta}}$$

As $\epsilon \to 0$, $\int_{C_\epsilon} f(z) \to 0$.
We know the pole of the function $f(z)$ is given by $z = -a$ which is inside the closed contour. Hence, the residue at $z = -a$ is given by

$$R_1 = \lim_{z \to -a} z^{-k} = (-a)^{-k} = a^{-k} e^{-ik\pi}$$

Thus,

$$\int_\epsilon^R f(z)dz + \int_R^\epsilon f(z)dz = 2\pi i a^{-k} e^{-ik\pi}$$

It is worth mentioning here that along $\overrightarrow{AB}$, $z = x$ such that $dz = dx$. However, along $\overrightarrow{DE}$, $z = x e^{2\pi i}$ and $dz = dx$. Hence when R tends to infinity and ϵ goes to zero, we have,

$$\int_0^\infty \frac{x^{-k}}{x+a}dx - \int_0^\infty \frac{x^{-k}}{x+a}e^{-2k\pi i}dx \;=\; 2\pi i a^{-k}e^{-ik\pi}$$

Simplifying the above equation we have

$$(1 - e^{-2k\pi i})(\int_0^\infty \frac{x^{-k}}{x+a}dx) \;=\; 2\pi i a^{-k}e^{-ik\pi}$$

Therefore,

$$\int_0^\infty \frac{x^{-k}}{x+a}dx \;=\; \frac{\pi a^{-k}}{\sin k\pi}$$

Example 11

Show that

$$\int_0^\infty \frac{xe^{-ax}}{x^2+b^2}dx = \frac{\pi}{2}\sin ab - \sin ab Si(ab) - \cos ab Ci(ab)$$

where Si and Ci are the sine and cosine integral functions respectively. This integral is important in the study of water waves mechanics which was first introduced by Sir Thomas Henry Havelock (1877-1968).

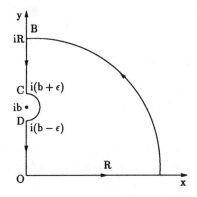

Fig. 4.9: A contour in the first quadrant.

Solution:

A suitable contour is shown in Fig. 4.9. Here the complex function is given by

$$f(z) = \frac{ze^{-az}}{z^2+b^2}$$

such that the poles are at $z = \pm ib$. Let us consider the contour

$$\Gamma = \overline{OA} + \widehat{AB} + \overline{BC} + \widehat{CD} + \overline{DO}$$

Now by the residue theorem $\int_\Gamma f(z)dz = 0$. Let the radius of the bigger circular arc be R and that of the small semicircular arc be ϵ. In the limit $R \to \infty$, $\epsilon \to 0$.

Along OA $\to$	$z = x$;	$dz = dx$
Along AB $\to$	$z = Re^{i\theta}$;	$dz = Rie^{i\theta}d\theta$
Along BC $\to$	$z = iy$;	$dz = idy$
Along CD $\to$	$z = ib + \epsilon e^{i\theta}$;	$dz = \epsilon i e^{i\theta}d\theta$
Along DO $\to$	$z = iy$;	$dz = idy$

Therefore,

$$\int_0^R \frac{xe^{-ax}}{x^2 + b^2}dx + \int_{AB} \frac{ze^{-az}}{z^2 + b^2}dz + \int_R^{b+\epsilon} \frac{iye^{-iay}}{b^2 - y^2}idy$$

$$+ \int_{CD} \frac{ze^{-az}}{z^2 + b^2}dz + \int_{b-\epsilon}^0 \frac{iye^{-iay}}{b^2 - y^2}idy = 0$$

or

$$I_1 + I_2 + I_3 + I_4 + I_5 = 0$$

As $\epsilon \to 0, R \to \infty$, we obtain

$$I_1 = \int_0^\infty \frac{xe^{-ax}}{x^2 + b^2}dx$$

Along the smaller circular contour CD

$$
\begin{aligned}
I_4 &= \int_{\pi/2}^{-\pi/2} \frac{(ib + \epsilon e^{i\theta})e^{-a(ib+\epsilon e^{i\theta})}}{(ib + \epsilon e^{i\theta})^2 + b^2} i\epsilon e^{i\theta}d\theta \\
&= \int_{\pi/2}^{-\pi/2} \frac{ibe^{-iab}}{2\epsilon ibe^{i\theta} + \epsilon^2 e^{2i\theta}} i\epsilon e^{i\theta}d\theta \\
&= \frac{-b}{2bi} \int_{\pi/2}^{-\pi/2} e^{-iab}d\theta \\
&= \frac{bi}{2b}e^{-iab}(-\pi) \\
&= \frac{-\pi i}{2}(\cos ab - i\sin ab)
\end{aligned}
$$

Therefore real part $= -(\pi/2)\sin ab$ and hence we obtain

$$I_4 = -\frac{\pi}{2}\sin ab$$

$$I_3 = \int_R^{b+\epsilon} \frac{iye^{-iay}}{b^2-y^2} idy = \int_\infty^b \frac{y}{y^2-b^2}e^{-iay}dy$$

$$= \frac{1}{2}\int_\infty^b \left(\frac{e^{-iay}}{y-b}+\frac{e^{-iay}}{y+b}\right)dy$$

$$= \frac{1}{2}\left[\int_\infty^0 \frac{e^{-ia(b+t)}}{t}dt + \int_\infty^{2b} \frac{e^{-ia(t-b)}}{t}dt\right]$$

$$2I_3 = e^{-iab}\int_\infty^0 \frac{\cos at - i\sin at}{t}dt + e^{iab}\int_\infty^{2b} \frac{\cos at - i\sin at}{t}dt$$

$$= (\cos ab - i\sin ab)\int_\infty^0 \frac{\cos at - i\sin at}{t}dt$$

$$+(\cos ab + i\sin ab)\int_\infty^{2b} \frac{\cos at - i\sin at}{t}dt$$

$$2\Re(I_3) = \cos ab\int_\infty^0 \frac{\cos at}{t}dt - \sin ab\int_\infty^0 \frac{\sin at}{t}dt$$

$$+ \cos ab\int_\infty^{2b} \frac{\cos at}{t}dt + \sin ab\int_\infty^{2b} \frac{\sin at}{t}dt$$

$I_2 \to 0$ as $R \to \infty$.

$$I_5 = \int_b^0 \frac{be^{-iay}}{y^2-b^2}dy = \frac{1}{2}\int_b^0 \left(\frac{e^{-iay}}{y-b}dt + \frac{e^{-iay}}{y+b}\right)dy$$

$$2I_5 = \int_0^{-b} \frac{e^{-ia(b+t)}}{t}dt + \int_{2b}^b \frac{e^{-ia(t-b)}}{t}dt$$

$$= (\cos ab - i\sin ab)\int_0^{-b} \frac{\cos at - i\sin at}{t}dt$$

$$+(\cos ab + i\sin ab)\int_{2b}^b \frac{\cos at - i\sin at}{t}dt$$

$$2\Re(I_5) = \cos ab\int_0^{-b} \frac{\cos at}{t}dt - \sin ab\int_0^{-b} \frac{\sin at}{t}dt$$

$$+ \cos ab\int_{2b}^b \frac{\cos at}{t}dt + \sin ab\int_{2b}^b \frac{\sin at}{t}dt$$

Therefore the value of the given integral is

$$I = \int_0^\infty \frac{xe^{-ax}}{x^2+b^2}dx = -I_3 - I_4 - I_5$$

$$I = \frac{\pi}{2}\sin ab + \frac{1}{2}\left[-\cos ab \int_\infty^0 \frac{\cos at}{t}dt + \sin ab \int_\infty^0 \frac{\sin at}{t}dt\right.$$

$$-\cos ab \int_\infty^{2b} \frac{\cos at}{t}dt - \sin ab \int_\infty^{2b} \frac{\sin at}{t}dt - \cos ab \int_0^{-b} \frac{\cos at}{t}dt$$

$$\left. + \sin ab \int_0^{-b} \frac{\sin at}{t}dt - \cos ab \int_{2b}^b \frac{\cos at}{t}dt - \sin ab \int_{2b}^b \frac{\sin at}{t}dt\right]$$

Combining the cosine terms yields

$$I_c = \frac{1}{2}\cos ab\left[\int_0^\infty \frac{\cos at}{t}dt + \int_{2b}^\infty \frac{\cos at}{t}dt + \int_{-b}^0 \frac{\cos at}{t}dt + \int_b^{2b} \frac{\cos at}{t}dt\right]$$

$$= \frac{1}{2}\cos ab\left[\int_0^\infty \frac{\cos at}{t}dt + \int_{2b}^\infty \frac{\cos at}{t}dt - \int_0^b \frac{\cos at}{t}dt\right.$$

$$\left. + \int_b^{2b} \frac{\cos at}{t}dt + \int_0^b \frac{\cos at}{t}dt - \int_0^b \frac{\cos at}{t}dt\right]$$

$$= \frac{1}{2}\cos ab \times 2\left[\int_0^\infty \frac{\cos at}{t}dt - \int_0^b \frac{\cos at}{t}dt\right]$$

$$= \cos ab \int_b^\infty \frac{\cos at}{t}dt = -\cos ab \int_\infty^b \frac{\cos at}{t}dt$$

$$= -Ci(ab)\cos ab$$

Here $Ci(x) = \int_\infty^x (\cos u/u)du$. Similarly collecting the sine terms we obtain

$$I_s = \frac{\sin ab}{2}\left[\int_\infty^0 + \int_{2b}^\infty + \int_0^{-b} + \int_b^{2b}\right]$$

$$= \frac{\sin ab}{2}\left(-\int_0^\infty + \int_{2b}^\infty - \int_0^b + \int_b^{2b} + \int_0^b - \int_0^b\right)$$

$$= \frac{\sin ab}{2}\left(-2\int_0^b\right)$$

$$= -\sin ab \int_0^b \frac{\sin at}{t}dt$$

$$= -Si(ab)\sin ab$$

Here $Si(x) = \int_0^\infty \frac{\sin u}{u}du$. Therefore

$$I = -Ci(ab)\cos ab - Si(ab)\sin ab + \frac{\pi}{2}\sin ab$$

Exercises

Evaluate the following integrals by the method of residues:

1. $\displaystyle\int_0^\infty \frac{dx}{1+x^2}$

2. $\displaystyle\int_0^\infty \frac{x\,dx}{(1+x^2)^2}$

3. $\displaystyle\int_0^\infty \frac{dx}{1+x^4}$

4. $\displaystyle\int_{-\infty}^\infty \frac{dx}{x^2+4}$

5. $\displaystyle\int_0^\infty \frac{\cos bx}{a^2+x^2}\,dx$

6. $\displaystyle\int_0^\infty \frac{\sin bx}{a^2+x^2}\,dx$

7. $\displaystyle\int_0^\infty \frac{x^2\,dx}{(a^2+x^2)(b^2+x^2)}$

8. $\displaystyle\int_0^\infty \frac{\cos bx}{(a^2+x^2)^2}\,dx$

9. $\displaystyle\int_0^\infty \frac{\cos \alpha x}{(x^2+a^2)(x^2+b^2)}\,dx$

10. $\displaystyle\int_{-\infty}^\infty \frac{\sin \alpha x}{(x^2+a^2)(x^2+b^2)}\,dx$

11. $\displaystyle\int_{-\infty}^\infty \frac{x\sin \alpha x}{(x^2+a^2)(x^2+b^2)}\,dx$

12. $\displaystyle\int_{-\infty}^\infty \frac{x\sin \alpha x}{a^4+x^4}\,dx$

13. $\displaystyle\int_0^\infty \frac{x^{p-1}}{(x+a)(x+b)}\,dx \qquad 0<p<2,\ \ 0<a,b$

14. $\displaystyle\int_0^\infty \frac{x^{p-1}}{(x+a)(x+b)(x+c)}\,dx \qquad 0<p<3,\ \ 0<a,b,c$

Hints: Use the full circular contour of very large radius indented to the right of the origin so that all the poles lie on the negative real axis to evaluate exercises 13 and 14

15. $\displaystyle\int_0^\infty \frac{x^{p-1}}{a^3+x^3}\,dx \qquad 0<p<3$

16. $\displaystyle\int_0^\infty \frac{x^{p-1}}{a^4+x^4}\,dx \qquad 0<p<4$

17. $\displaystyle\int_0^{2\pi} \frac{1 - 2x\cos\theta}{x^2 + 1 - 2x\cos\theta}\,d\theta \qquad -1 < x < 1$

18. $\displaystyle\int_0^{2\pi} \frac{d\theta}{5 + 4\sin\theta}$

19. $\displaystyle\int_0^{2\pi} \frac{d\theta}{(a + b\cos\theta)^2} \qquad 0 < b < a$

20. $\displaystyle\int_0^{2\pi} \frac{\cos 3\theta}{5 + 3\sin\theta}\,d\theta$

21. $\displaystyle\int_0^{2\pi} \frac{\cos^2\theta\,d\theta}{a + b\cos\theta} \qquad 0 < b < a$

22. $\displaystyle\int_0^{\pi} \frac{\cos 2\theta\,d\theta}{5 + 4\cos\theta}$

Use residues to establish the following formulas:

23. $\displaystyle\int_0^{\infty} \frac{dx}{1 + x^4} = \frac{\pi}{2\sqrt{2}}$

24. $\displaystyle\int_0^{\pi} \frac{\cos^2 3\theta}{5 - 4\cos 2\theta}\,d\theta = \frac{3\pi}{8}$

25. $\displaystyle\int_0^{\pi} \frac{\cos 2\theta\,d\theta}{1 - 2\alpha\cos\theta + \alpha^2} = \frac{\pi\alpha^2}{1 - \alpha^2} \qquad -1 < \alpha < 1$

26. $\displaystyle\int_0^{\pi} \frac{d\theta}{(\alpha + \cos\theta)^2} = \frac{\pi\alpha}{(\alpha^2 - 1)^{3/2}} \qquad \alpha > 1$

27. $\displaystyle\int_0^{\pi} \sin^{2n}\theta\,d\theta = \frac{(2n)!}{2^{2n}(n!)^2}\pi \qquad n = 1, 2, \cdots$

28. $\displaystyle\int_0^{\infty} \frac{\sin^2 x\,dx}{x^2} = \frac{\pi}{2}$

29. $\displaystyle\int_{-\infty}^{\infty} \frac{dx}{(x^2 + 4x + 5)^2} = \frac{\pi}{2}$

30. $\displaystyle\int_0^{\infty} \frac{\cos 2\pi x}{x^4 + x^2 + 1}\,dx = \frac{-\pi}{2\sqrt{3}}e^{-\pi\sqrt{3}}$

31. Show that
$$\int_0^{\infty} \frac{\sin x}{x}\,dx = \frac{\pi}{2}$$

Hint: Integrate $\dfrac{e^{iz}}{z}$ around the contour shown in Fig. 4.10, and let $\epsilon \to 0$ and $R \to \infty$.

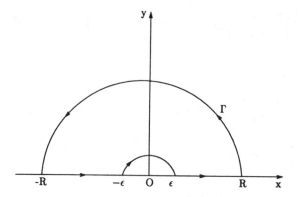

Fig. 4.10: A semi-circular contour.

32. Using the semi-circular contour in Fig. 4.10, show that

(a) $\displaystyle\int_0^\infty \frac{(\ln x)^2}{1+x^2}\,dx = \frac{\pi^3}{8}$

(b) $\displaystyle\int_0^\infty \frac{\ln x}{1+x^2}\,dx = 0$

33. Evaluate the following integrals:

(a) $\displaystyle\int_0^\infty \frac{(\ln x)^2}{x^2+b^2}\,dx$

(b) $\displaystyle\int_0^\infty \frac{\ln x}{b^2+x^2}\,dx$

34. The Fresnel integrals

$$\int_0^\infty \cos(x^2)\,dx = \int_0^\infty \sin(x^2)\,dx = \frac{\sqrt{2\pi}}{4}$$

are important in diffraction theory. It is defined that

$$\mathrm{erf}(x) = \frac{2}{\sqrt{\pi}} \int_0^x e^{-\eta^2}\,d\eta$$

is an error function whose value at $x = \infty$ is unity. Evaluate these integrals by integrating $\exp(i\,z^2)$ around the boundary of the sector $0 \le r \le R,\, 0 \le \theta \le \pi/4$ and then letting $R \to \infty$ (see Fig. 4.11).

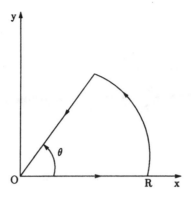

Fig. 4.11: A sector contour.

35. Show that

$$\int_0^\infty \frac{\cos x}{\sqrt{x}}\,dx = \int_0^\infty \frac{\sin x}{\sqrt{x}}\,dx = \sqrt{\frac{\pi}{2}}$$

Hint: Integrate $\dfrac{e^{iz}}{\sqrt{z}}$ around the contour shown in Fig. 4.12, let $\epsilon \to 0$ and $R \to \infty$ and use the formula $\displaystyle\int_0^\infty e^{-\eta^2}\,d\eta = \dfrac{\sqrt{\pi}}{2}$.

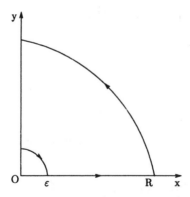

Fig. 4.12: A circular contour in the 1st quadrant.

36. Show that

$$\int_{-\infty}^{\infty} \frac{\cos \alpha x}{e^x + e^{-x}} dx = \frac{\pi}{e^{\alpha\pi/2} + e^{-\alpha\pi/2}}$$

Hint: Integrate the function $\dfrac{e^{i\alpha z}}{e^z + e^{-z}}$ around the contour shown in Fig. 4.13, and let $R \to \infty$.

37. The function

$$\beta(m,n) = \int_0^1 x^{m-1}(1-x)^{n-1} dx, \quad m > 0, \quad n > 0$$

is defined as **beta function**. By using a suitable contour show that

$$\beta(m, 1-m) = \frac{\pi}{\sin m\pi} \quad 0 < m < 1$$

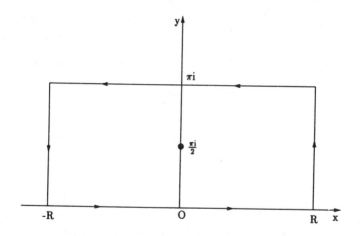

Fig. 4.13: A rectangular contour.

38. Show that
 (a) $\int_{-\infty}^{\infty} \{x^2/(1+x^4)^2\} dx = \pi/(4\sqrt{2})$
 (b) $\int_0^{2\pi} \{\sin^2 \theta/(a + b\cos \theta)\} d\theta = (2\pi/b^2)(a - \sqrt{a^2 - b^2}) \quad 0 < b < a$

39. Show that
 $\int_0^{\infty} \{\sin x/(1+x^2)\} dx = \sinh(1)Ei(1) + e^{-1}shi(1)$ where Ei and shi are the exponential and hyperbolic integrals respectively.

Chapter 5

The theory of transform calculus and applications

5.1 Fourier series

5.1.1 Introduction

In the eighteenth century, Jean B.J. Fourier (1768-1830), a French mathematician and physicist, a confidant of Napoleon first discovered in the heat conduction research that a periodic function can be expressed in an infinite series of sine and cosine terms if it is properly defined. This discovery led to the introduction of Fourier series for solving problems of physical interest.

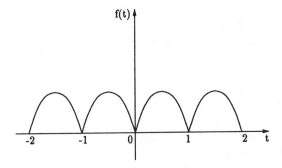

Fig. 5.1: A typical periodic function.

A function $f(t)$ is periodic if, and only if, there exists a positive parameter $2p$ such that for every t in $[0, \infty)$, $f(t + 2p) = f(t)$. The parameter $2p$ is called a period of $f(t)$.

Thus with this definition we can show that if $f(t)$ is periodic, then

$$f(t) = f(t + 2p) = f(t + 4p) = f(t + 6p) = f(t + 2np) = \cdots$$

where $n = 1, 2, 3, \cdots$ and hence $2p, 4p, 6p, 8p, \cdots, 2np$ are also periods of f. Here $2p$ is the smallest of all the periods and is usually defined as the fundamental period of f. A typical periodic function is depicted in Fig. 5.1.

5.1.2   Fourier series and Fourier coefficients

Let $f(t)$ satisfy the following conditions:

(a) $f(t)$ is defined in the interval $d < t < d + 2p$, where d is a real number.

(b) $f(t)$ and $f'(t)$ are sectionally continuous in the interval $d < t < d + 2p$.

(c) $f(t + 2p) = f(t)$, i.e. $f(t)$ is periodic with period $2p$.

Then at every point of continuity, $f(t)$ can be expressed as an infinite series of cosine and sine, and is given by

$$f(t) = \frac{a_0}{2} + \sum_{n=1}^{\infty} \left\{ a_n \cos \frac{n\pi t}{p} + b_n \sin \frac{n\pi t}{p} \right\} \tag{5.1}$$

where

$$a_n = \frac{1}{p} \int_d^{d+2p} f(t) \cos \frac{n\pi t}{p} dt \qquad n = 0, 1, 2, 3, \cdots \tag{5.2}$$

$$b_n = \frac{1}{p} \int_d^{d+2p} f(t) \sin \frac{n\pi t}{p} dt \qquad n = 1, 2, 3, \cdots \tag{5.3}$$

Here (5.1) is known as Fourier series and a_n and b_n are called Fourier coefficients.

At a point of discontinuity, the left-hand side of (5.1) should be modified to yield

$$f(t) = \frac{1}{2}[f(t+0) + f(t-0)] \tag{5.4}$$

The above conditions (a), (b) and (c) together are known as the Dirichlet conditions and are sufficient (but not necessary) conditions for convergence of Fourier series. Thus any periodic function with fundamental period $2p$ can be expressed as in (5.1) with the assumption that the infinite series converges.

The Fourier series of a function $f(t)$ is defined whenever all the integrals in (5.2) and (5.3) have meaning. This is certainly the case if $f(t)$ is continuous on the interval $d \le t \le d + 2p$. However, the integrals also have meaning when $f(t)$ has jump discontinuities, as shown in Fig. 5.2.

The mathematical definition of the given periodic function is:

$$f(t) = \begin{cases} f_1(t) & d < t < t_1 \\ f_2(t) & t_1 < t < t_2 \\ f_3(t) & t_2 < t < d + 2p \end{cases}$$

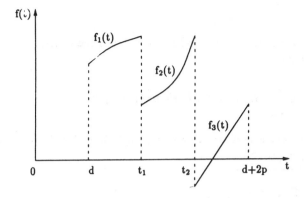

Fig. 5.2: A piecewise continuous periodic function.

This type of function is defined as the piecewise continuous function. Hence it is necessary to write the Fourier coefficients as

$$
\begin{aligned}
a_n &= \frac{1}{p} \int_d^{d+2p} f(t) \cos \frac{n\pi t}{p} dt \\
&= \frac{1}{p} \int_d^{t_1} f_1(t) \frac{n\pi t}{p} dt + \frac{1}{p} \int_{t_1}^{t_2} f_2(t) \cos \frac{n\pi t}{p} dt \\
&\quad + \frac{1}{p} \int_{t_1}^{d+2p} f_3(t) \cos \frac{n\pi t}{p} dt
\end{aligned}
\tag{5.5}
$$

$$
\begin{aligned}
b_n &= \frac{1}{p} \int_d^{d+2p} f(t) \sin \frac{n\pi t}{p} dt \\
&= \frac{1}{p} \int_d^{t_1} f_1(t) \sin \frac{n\pi t}{p} dt + \frac{1}{p} \int_{t_1}^{t_2} f_2(t) \sin \frac{n\pi t}{p} dt \\
&\quad \frac{1}{p} \int_{t_2}^{d+2p} f_3(t) \sin \frac{n\pi t}{p} dt
\end{aligned}
\tag{5.6}
$$

At the jump point it is not obvious what value to assign to f. We shall always use the average of left and right limits given by (5.4).

Example 1

Find the Fourier series expansion of the periodic function whose definition in one period is $f(t) = t \qquad -1 < t < 1$.

Solution

In this case $2p = 2$ so that $p = 1$ and $d = -1$. Thus the Fourier series for $f(t)$ exists in the following form:

$$f(t) = \frac{a_0}{2} + \sum_{n=1}^{\infty} \{a_n \cos n\pi t + b_n \sin n\pi t\}$$

where the Fourier coefficients are obtained using (5.2) and (5.3). Thus we have $a_n = 0$, for all n and $b_n = (2/n\pi)(-1)^{n+1}$.

Hence the Fourier series is

$$f(t) = \frac{2}{\pi} \sum_{n=1}^{\infty} \frac{(-1)^{n+1} \sin n\pi t}{n}$$

5.1.3   Complex form of Fourier series

Another useful form of Fourier Series is the complex exponential form. The form is obtained by substituting the exponential equivalent of the cosine and sine terms into the original form of Fourier series:

$$
\begin{aligned}
f(t) &= \frac{a_0}{2} + \sum_{n=1}^{\infty} \left\{ a_n \cos \frac{n\pi t}{p} + b_n \sin \frac{n\pi t}{p} \right\} \\
&= \frac{a_0}{2} + \sum_{n=1}^{\infty} \left\{ a_n \left(\frac{e^{ni\pi t/p} + e^{-ni\pi t/p}}{2} \right) + b_n \left(\frac{e^{ni\pi t/p} - e^{-ni\pi t/p}}{2i} \right) \right\} \quad (5.7)
\end{aligned}
$$

Collecting the coefficients of $e^{ni\pi t/p}$ and $e^{-ni\pi t/p}$, we obtain

$$f(t) = \frac{a_0}{2} + \sum_{n=1}^{\infty} \left(\frac{a_n - ib_n}{2} \right) e^{ni\pi t/p} + \sum_{n=1}^{\infty} \left(\frac{a_n + ib_n}{2} \right) e^{-ni\pi t/p} \quad (5.8)$$

We can combine these expressions into one series if we define

$$c_0 = \frac{a_0}{2}, \quad c_n = \frac{a_n - ib_n}{2}, \quad c_{-n} = \frac{a_n + ib_n}{2} = c_n^*$$

such that the last series can be written in more symmetric form

$$f(t) = \sum_{n=-\infty}^{\infty} c_n e^{ni\pi t/p} \quad (5.9)$$

Here it is noted that $c_{-n} = c_n^*$ is the complex conjugate of c_n. This series is known as the complex form of the Fourier series and represents a real quantity. The coefficients

c_0, c_n and c_{-n} can be calculated directly from their definitions. Thus,

$$c_0 = \frac{a_0}{2} = \frac{1}{2p} \int_d^{d+2p} f(t)dt$$

$$c_n = \frac{1}{2}(a_n - ib_n) = \frac{1}{2p} \int_d^{d+2p} f(t) \left\{ \cos \frac{n\pi t}{p} - i \sin \frac{n\pi t}{p} \right\} dt$$

$$= \frac{1}{2p} \int_d^{d+2p} f(t)e^{-ni\pi t/p} dt$$

$$c_{-n} = \frac{1}{2}(a_n + ib_n) = \frac{1}{2p} \int_d^{d+2p} f(t)e^{ni\pi t/p} dt$$

Obviously, the coefficient c_n can be calculated from the single formula

$$c_n = \frac{1}{2p} \int_d^{d+2p} f(t)e^{-ni\pi t/p} dt, \quad n = 0, \pm 1, \pm 2, \cdots \tag{5.10}$$

Example 2

Find the complex form of the Fourier series of the periodic function whose definition in one period is:

$$f(t) = \begin{cases} 1 & 0 < t < 1 \\ 0 & 1 < t < 2 \end{cases}$$

Solution

In this case $2p = 2$ so that $p = 1$ and $d = 0$. The complex form of Fourier series is given by

$$f(t) = \sum_{n=-\infty}^{\infty} c_n e^{ni\pi t/p} = \sum_{n=-\infty}^{\infty} c_n e^{ni\pi t}$$

The complex Fourier coefficients are given by $c_n = (1/2ni\pi)\{1 - e^{-ni\pi}\}$ and $c_0 = 1/2$. Hence the complex series is

$$f(t) = \frac{1}{2} + \sum_{-\infty}^{\infty 1} \left(\frac{1 - e^{-ni\pi}}{2ni\pi} \right) e^{ni\pi t}$$

where $\sum^1$ denotes a sum excluding the value for $n = 0$.

Now the above series can be converted to trigonometric series as follows, since we know $c_n = (a_n - ib_n)/2$ and $c_{-n} = (a_n + ib_n)/2$. Thus $a_n = c_n + c_{-n}$ and $b_n = i(c_n - c_{-n})$.

After reduction we obtain

$$a_n = 0$$

$$b_n = \begin{cases} 2/n\pi & n = 1, 3, 5, \cdots \\ 0 & n = 2, 4, 6, \cdots \end{cases}$$

Hence, we can write

$$f(t) = \frac{1}{2} + \frac{2}{\pi} \sum_{n=1,3,5,\cdots}^{\infty} \frac{\sin n\pi t}{n}$$

5.1.4   Half-range Fourier sine and cosine series

In many of the applications of Fourier series it may be noticed that $f(t)$ possesses certain symmetry properties. The given function may be symmetric in the origin or it may be symmetric about the y-axis. In that situation, evaluations of the Fourier coefficients are extremely simple. When the function is symmetric in the origin, i.e. when it is an odd function, the Fourier expansion contains only sine terms; on the other hand, when the function is symmetric about the y-axis, i.e. when it is an even function, the Fourier expansion contains only cosine terms.

Suppose $f(t)$ is periodic with period $2p$ and is defined in $(-p, p)$. Then the Fourier series for the even and odd periodic functions with their coefficients can be written, respectively, as

$$f(t) = \frac{a_0}{2} + \sum_{n=1}^{\infty} a_n \cos \frac{n\pi t}{p}$$

$$a_n = \frac{2}{p} \int_0^p f(t) \cos \frac{n\pi t}{p} dt \tag{5.11}$$

$$f(t) = \sum_{n=1}^{\infty} b_n \sin \frac{n\pi t}{p}$$

$$b_n = \frac{2}{p} \int_0^p f(t) \sin \frac{n\pi t}{p} dt \tag{5.12}$$

It is to be noted here that this observation of even and odd functions of a periodic function $f(t)$ reduces the labour of calculating the Fourier coefficients by half.

From the integrals in eqns (5.11) and (5.12), it may be observed as if the given periodic function with period $2p$ is defined in the interval $[0, p]$. In practice, in many of the applications of Fourier series a function $f(t)$ is defined on the interval $0 < t < p$. We may then represent this function either as a series consisting of only cosine terms (5.11), or as one consisting of only sine terms (5.12). This is accomplished by making either an even or an odd periodic extension of the given function, respectively.

The Fourier series expansions of (5.11) and (5.12) are respectively called the half-range Fourier cosine and sine series. The concept of even and odd extension of a periodic function is demonstrated below with examples.

Example 3

Find the Fourier series expansion of the periodic function whose definition in one period is

$$f(t) = t \qquad -1 < t < 1$$

Solution

We see that $f(t)$ is defined in $-1 < t < 1$. Then, $f(-t) = -t = -f(t)$ and thus the function is periodic with period $2p = 2$ so that $p = 1$.

The Fourier series is $f(t) = \sum_{n=1}^{\infty} b_n \sin n\pi t$ where the coefficients are simply $b_n = (2/n\pi)(-1)^{n+1}$. Therefore, the Fourier series is

$$f(t) = \frac{2}{\pi} \sum_{n=1}^{\infty} (-1)^{n+1} \frac{\sin n\pi t}{n}$$

Example 4

Find the Fourier cosine series expansion of the function $f(t) = t,\ 0 < t < 1$.

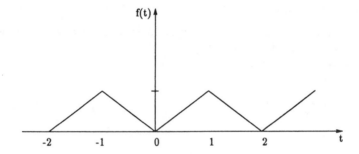

Fig. 5.3: Even periodic extension.

Solution

The even periodic expansion of the function is graphed in the following, Fig. 5.3.

Here the period $= 2p = 2$, therefore $p = 1$. The Fourier series is

$$f(t) = \frac{a_0}{2} + \sum_{n=1}^{\infty} a_n \cos n\pi t$$

where the coefficients are simply

$$
\begin{aligned}
a_n &= \frac{2}{n^2\pi^2}[(-1)^n - 1] \\
&= \begin{cases} 0 & \text{if } n \text{ is even} \quad n \neq 0 \\ -4/(n^2\pi^2) & \text{if } n \text{ is odd} \end{cases} \\
a_0 &= 2\int_0^1 t\, dt = \frac{2}{2} = 1
\end{aligned}
$$

Hence the required series can be written as

$$f(t) = \frac{1}{2} - \frac{4}{\pi^2} \sum_{n=1,3,5,\cdots}^{\infty} \frac{\cos n\pi t}{n^2}$$

5.1.5   Parseval's identity

Assuming that the Fourier series of $f(t)$ converges uniformly in $(-p, p)$ then Parseval's identity can be written as

$$\frac{1}{p} \int_{-p}^{p} \{f(t)\}^2 dt = \frac{c_0^2}{2} + \sum_{n=1}^{\infty}(a_n^2 + b_n^2)$$

where the integral is assumed to exist.

This identity can be used now to determine the electric power delivered by an electric current, $I(t)$, flowing under a voltage, $E(t)$, through a resistor of resistance, R.

From the theory of electricity, it is known that this power is given by

$$P = EI$$
$$= RI^2, \quad \text{since } E = RI$$

In most applications the current $I(t)$ is a periodic function and it is important to know the average power delivered per cycle.

Thus, the average power is

$$P_{av} = \frac{1}{2p} \int_{-p}^{p} RI^2(t)dt$$

$$= \frac{R}{2p} \int_{-p}^{p} I^2(t)dt \qquad (5.13)$$

Suppose that $I(t)$ is given as a Fourier series

$$I(t) = \frac{a_0}{2} + \sum_{n=1}^{\infty}\left[a_n \cos \frac{n\pi t}{p} + b_n \sin \frac{n\pi t}{p}\right]$$

Then using Parseval's identity, we have

$$P_{av} = \frac{R}{2}\left[\frac{a_0^2}{2} + \sum_{n=1}^{\infty}(a_n^2 + b_n^2)\right]$$

$$= R\left[\frac{a_0^2}{4} + \frac{1}{2}\sum_{n=1}^{\infty}(a_n^2 + b_n^2)\right] \qquad (5.14)$$

But mean square of the current $I(t)$ is given by

$$I_{av} = \frac{1}{2p} \int_{-p}^{p} I^2(t)dt$$

$$= \frac{a_0^2}{4} + \frac{1}{2}\sum_{n=1}^{\infty}(a_n^2 + b_n^2)$$

Therefore, the root mean-square of the current is

$$I_{\text{rms}} = \sqrt{\frac{a_0^2}{4} + \frac{1}{2}\sum_{n=1}^{\infty}(a_n^2 + b_n^2)} \qquad (5.15)$$

which is usually denoted by I_{rms}.

Exercises

1. Find the Fourier expansion of the periodic function whose definition in one period is

   $$f(t) = 4 - t^2 \quad -2 \le t \le 2$$

   Using this result, establish the following:

   (a) $1 + 1/2^2 + 1/3^2 + 1/4^2 + \cdots = \pi^2/6$
   (b) $1 + 1/3^2 + 1/5^2 + 1/7^2 + \cdots = \pi^2/8$

2. Find the Fourier series expansion of the periodic function whose definition in one period is

   $$f(t) = t^2, \quad 0 < t < 2\pi$$

   Using this result, obtain the following numerical result

   $$1 - \frac{1}{2^2} + \frac{1}{3^2} - \frac{1}{4^2} + \frac{1}{5^2} - \cdots = \frac{\pi^2}{12}$$

3. Find the complex forms of the Fourier series of the periodic functions whose definition in one period is:

   (a) $f(t) = e^t \quad -\pi < t < \pi$
   (b) Using this result, show that the real Fourier series is

   $$f(t) = \frac{2 \sinh \pi}{\pi} \left[\frac{1}{2} - \frac{1}{1+1^2}(\cos t - \sin t) + \frac{1}{1+2^2}(\cos 2t - 2 \sin 2t) \cdots \right]$$

4. $f(t) = \sin t \quad 0 \le t \le \pi$

5. Show that the amplitude of the harmonic in the Fourier series of a periodic function $f(t)$ is equal to $2\sqrt{c_n c_{-n}}$.

   Find the half-range sine and cosine expressions for each of the following functions:

6. (a) Expand $f(t) = \cos t$, $0 < t < \pi$, into a Fourier sine series.
   (b) How should $f(t)$ be defined at $t = 0$ and $t = \pi$ so that the series will converge to $f(t)$ for $0 \le t \le \pi$?

7. Expand the following periodic function $(f(t) = f(t + 8))$

   $$f(t) = \begin{cases} t & 0 < t < 4 \\ 8 - t & 4 < t < 8 \end{cases}$$

8. Express the I_{rms} in terms of the coefficients in the complex form of the Fourier expansion of $I(t)$.

9. If $f(t)$ and $g(t)$ are two periodic functions of period $2p$ defined over $(-p, p)$ and are expressed in complex form of Fourier expansions as

$$f(t) = \sum_{n=-\infty}^{\infty} c_n e^{in\pi t/p}$$

$$g(t) = \sum_{m=-\infty}^{\infty} d_m e^{im\pi t/p}$$

show that the average value of the product $f(t)g(t)$ over one period is

$$\sum_{n=-\infty}^{\infty} c_n d_{-n}$$

10. If $f(t)$ is the periodic function whose definition in one period is

$$f(t) = \begin{cases} 1 & 0 < t < 1 \\ -1 & 1 < t < 2 \end{cases}$$

find a solution for each of the following equations which satisfies the given initial conditions:

(a) $y'' - y = f(t) \qquad y_0 = y_0' = 0$

(b) $y'' + 3y' + 2y = f(t) \qquad y_0 = y_0' = 0$

5.2   Fourier integral and transform

We have already seen that a periodic function can adequately be represented by a Fourier series expansion satisfying the Dirichlet conditions. In previous sections, we have illustrated the responses of mechanical and electrical systems to general periodic disturbances. However, in many problems of physical interests, the impressed force on the applied voltage is nonperiodic rather than periodic. In that situation, Fourier series expansion cannot represent a nonperiodic function. The nonperiodic function may be obtained when the period of a periodic function goes to infinity. Under this limiting condition, a Fourier series tends to Fourier integral. For rigorous mathematical treatment of the subject, the reader is directed to the book of I.N. Sneedon , *Fourier Transforms* (McGraw-Hill, New York, 1951).

The following problem will illustrate how a periodic function becomes a nonperiodic function when the period goes to infinity.

For simplicity, consider the following example whose definition in one period is

$$f(t) = e^{-|t|} \qquad -p < t < p \tag{5.16}$$

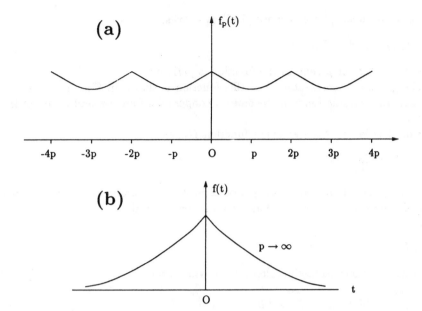

Fig. 5.4: (a) Sketch of periodic behaviour of function $f(t)$; (b) Sketch of nonperiodic behaviour of function $f(t)$.

This periodic function can be represented in Fourier series

$$f(t) = \frac{a_0}{2} + \sum_{n=1}^{\infty} \left(a_n \cos \frac{n\pi t}{p} + b_n \sin \frac{n\pi t}{p} \right) \qquad (5.17)$$

where

$$a_n = \frac{1}{p} \int_{-p}^{p} f(t) \cos \frac{n\pi t}{p} dt$$

$$b_n = \frac{1}{p} \int_{-p}^{p} f(t) \sin \frac{n\pi t}{p} dt$$

Since the given function (5.16) is an even periodic function, we would have $b_n = 0$ for all n. Hence, we have

$$f(t) = \frac{a_0}{2} + \sum_{n=1}^{\infty} a_n \cos \frac{n\pi t}{p} \qquad (5.18)$$

where

$$a_n = \frac{2}{p} \int_{0}^{p} f(t) \cos \frac{n\pi t}{p} dt$$

For our convenience, let us denote (5.16) as follows:

$$f_p(t) = e^{-|t|} \qquad -p < t < p$$

where the subscript p denotes the function is periodic.

The graphical representation of this function is shown in Fig. 5.4(a) and when $p \to \infty$, this periodic function becomes a nonperiodic function and is shown in Fig. 5.4(b).

This then leads us to define the function $f(t)$ as

$$f(t) = \lim_{p \to \infty} f_p(t)$$

Now the $f(t)$ function is no longer periodic. Further we assume that $f(t)$ is absolutely integrable on the real line, i.e. that the improper integral

$$\int_{-\infty}^{\infty} |f(t)| dt$$

be finite. From our sample problem, it is obvious that

$$\int_{-\infty}^{\infty} |e^{-|t|}| dt \;=\; 2 \int_{0}^{\infty} e^{-t} dt = 2 \lim_{p \to \infty} \int_{0}^{p} e^{-t} dt = 2$$

which is finite.

Now to establish the Fourier integral as the limit of a Fourier series, we begin with the complex exponential form of a Fourier series,

$$f_p(t) = \sum_{n=-\infty}^{\infty} c_n e^{n\pi i t/p} \tag{5.19}$$

where

$$c_n \;=\; \frac{1}{2p} \int_{-p}^{p} f_p(t) e^{-n\pi it/p} dt = \frac{1}{2p} \int_{-p}^{p} f_p(\tau) e^{-n\pi i\tau/p} d\tau \tag{5.20}$$

Substitution of the second expression of c_n into (5.19), yields,

$$
\begin{aligned}
f_p(t) \;&=\; \sum_{n=-\infty}^{\infty} \left[\frac{1}{2p} \int_{-p}^{p} f_p(\tau) e^{-n\pi i\tau/p} d\tau \right] e^{\frac{n\pi it}{p}} \\
&=\; \sum_{n=-\infty}^{\infty} \left[\frac{1}{2\pi} \int_{-p}^{p} f_p(\tau) e^{-n\pi i\tau/p} d\tau \right] e^{n\pi it/p} \left(\frac{\pi}{p} \right)
\end{aligned}
\tag{5.21}
$$

Define the frequency of the general term by $\sigma_n = n\pi/p$ and difference in frequency between successive terms by $\Delta\sigma = \sigma_{n+1} - \sigma_n = \{(n+1)\pi\}/p - n\pi/p = \pi/p$.

Then (5.21) reduces to

$$f_p(t) = \sum_{n=-\infty}^{\infty} \left[\frac{1}{2\pi} e^{i\sigma_n t} \int_{-p}^{p} f_p(\tau) e^{-i\sigma_n \tau} d\tau \right] \Delta\sigma \tag{5.22}$$

We know that as $p \to \infty$, $\Delta\sigma \to 0$. Thus as $p \to \infty$, eqn (5.22) can be written as an integral,

$$f(t) = \int_{-\infty}^{\infty} \left[\frac{1}{2\pi} e^{i\sigma t} \int_{-\infty}^{\infty} f(\tau) e^{-i\sigma\tau} d\tau \right] d\sigma \tag{5.23}$$

which is known as the complex form of Fourier integral.

Equation (5.23) is actually a valid representation of the nonperiodic limit function $f(t)$, provided that:

(a) In every finite interval $f(t)$ satisfies the Dirichlet conditions.

(b) The improper integral $\int_{-\infty}^{\infty} |f(t)| dt$ exists.

Complex Fourier transform

Referring to eqn (5.23), we are now in a position to define the complex Fourier transform pair.

Define

$$g(\sigma) = \mathcal{F}\{f(t)\} = \int_{-\infty}^{\infty} e^{-i\sigma t} f(t) dt \tag{5.24}$$

then

$$\begin{aligned} f(t) &= \mathcal{F}^{-1}\{g(\sigma)\} \\ &= \frac{1}{2\pi} \int_{-\infty}^{\infty} g(\sigma) e^{i\sigma t} d\sigma \end{aligned} \tag{5.25}$$

Relations (5.24) and (5.25) are together called Fourier transform pairs. Here $g(\sigma)$ is in the frequency domain and $f(t)$ in the time domain.

Fourier integrals

After a little bit of reduction, eqn (5.23) can be expressed as follows:

$$\begin{aligned} f(t) &= \frac{1}{\pi} \int_{0}^{\infty} \int_{-\infty}^{\infty} f(\tau) \cos\sigma\tau \cos\sigma t \, d\tau d\sigma \\ &\quad + \frac{1}{\pi} \int_{0}^{\infty} \int_{-\infty}^{\infty} f(\tau) \sin\sigma\tau \sin\sigma t \, d\tau d\sigma \end{aligned} \tag{5.26}$$

In analogy to the Fourier coefficients, if we define

$$a(\sigma) = \int_{-\infty}^{\infty} f(t) \cos\sigma t \, dt \tag{5.27}$$

$$b(\sigma) = \int_{-\infty}^{\infty} f(t) \sin\sigma t \, dt \tag{5.28}$$

Equation (5.26) can be rewritten as

$$f(t) = \frac{1}{\pi} \int_{0}^{\infty} a(\sigma) \cos\sigma t \, d\sigma + \frac{1}{\pi} \int_{0}^{\infty} b(\sigma) \sin\sigma t \, d\sigma \tag{5.29}$$

This equation is called the Fourier integral of $f(t)$.

Fourier cosine transform

Now it can be easily verified that if $f(t)$ is an even function of t then $b(\sigma) = 0$ and then we can define the Fourier cosine transform pair as follows.
   Define

$$a(\sigma) \;=\; \mathcal{F}\{f(t)\} = 2 \int_0^\infty f(t) \cos \sigma t \, dt \tag{5.30}$$

then

$$f(t) \;=\; \mathcal{F}^{-1}\{g(\sigma)\} = \frac{1}{\pi} \int_0^\infty g(\sigma) \cos \sigma t \, d\sigma \tag{5.31}$$

Relations (5.30) and (5.31) are called Fourier cosine transform pairs and this transform exists when $f(t)$ is an even function.

Fourier sine transform

Now it can be easily verified that if $f(t)$ is an odd function of t then $a(\sigma) = 0$ and then we can define the Fourier cosine transform pair as follows.
   Define

$$b(\sigma) \;=\; \mathcal{F}\{f(t)\} = 2 \int_0^\infty f(t) \sin \sigma t \, dt \tag{5.32}$$

then

$$f(t) \;=\; \mathcal{F}^{-1}\{g(\sigma)\} = \frac{1}{\pi} \int_0^\infty g(\sigma) \sin \sigma t \, d\sigma \tag{5.33}$$

These two relations are called Fourier sine transform pairs which exist only when $f(t)$ is an odd function.

Example 5

Find the Fourier integral representation of the following nonperiodic function.

$$f(t) = \begin{cases} \sin t & t^2 \le \pi^2 \\ 0 & t^2 \ge \pi^2 \end{cases}$$

Solution

To obtain the Fourier integral representation of this function, we first obtain the Fourier transform and then obtain the inverse of the transform. The graphical representation is in Fig. 5.5.

$$\begin{aligned}
\mathcal{F}\{f(t)\} &= \int_{-\infty}^\infty f(t) e^{-i\sigma t} \, dt \\
&= \int_{-\pi}^\pi \sin t\, e^{-i\sigma t} \, dt \\
&= -\frac{e^{i\sigma\pi} - e^{-i\sigma\pi}}{1 - \sigma^2} = \frac{-2i \sin \sigma \pi}{1 - \sigma^2}
\end{aligned}$$

Thus

$$f(t) = \mathcal{F}^{-1}\left[\frac{-2i \sin \sigma \pi}{1 - \sigma^2}\right]$$

$$= \frac{1}{2\pi}\int_{-\infty}^{\infty}\left[\frac{-2i \sin \sigma \pi}{1 - \sigma^2}\right]e^{i\sigma t}d\sigma$$

$$= -\frac{i}{\pi}\int_{-\infty}^{\infty}\frac{\sin \sigma \pi[\cos \sigma t + i \sin \sigma t]}{1 - \sigma^2}d\sigma$$

Therefore after a reduction

$$f(t) = \frac{2}{\pi}\int_0^{\infty}\frac{\sin \sigma \pi \sin \sigma t}{1 - \sigma^2}d\sigma$$

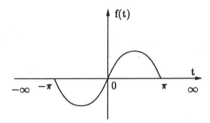

Fig. 5.5: Graphical representation of $f(t)$.

5.2.1 Relationship between Fourier and Laplace transforms

In using the applications of the Fourier integral the function to be represented is identically zero before some instant, usually $t = 0$. In that situation the general Fourier transform pair given by (5.24) and (5.25) reduce to,

$$g(\sigma) = \mathcal{F}\{f(t)\} = \int_0^{\infty} e^{-i\sigma t}f(t)dt \qquad (5.34)$$

where

$$f(t) = \frac{1}{2\pi}\int_{-\infty}^{\infty} g(\sigma)e^{i\sigma t}d\sigma \qquad (5.35)$$

Now consider the following function,

$$F(t) = \begin{cases} e^{-at}f(t) & t > 0 \\ & \qquad a > 0 \\ 0 & t < 0 \end{cases} \qquad (5.36)$$

where $f(t)$ is the function in which we are interested. Then taking the Fourier transform of $F(t)$, we obtain,

$$G(\sigma) = \mathcal{F}\{F(t)\} = \int_{-\infty}^{\infty} e^{-i\sigma t} F(t) dt \tag{5.37}$$

$$= \int_{0}^{\infty} e^{-i\sigma t} [e^{-at} f(t)] dt \tag{5.38}$$

Therefore,

$$G(\sigma) = \int_{0}^{\infty} e^{-(a+i\sigma)t} f(t) dt \tag{5.39}$$

whose inverse transform is

$$F(t) = e^{-at} f(t) = \frac{1}{2\pi} \int_{-\infty}^{\infty} G(\sigma) e^{i\sigma t} d\sigma$$

Therefore,

$$f(t) = \frac{1}{2\pi} \int_{-\infty}^{\infty} G(\sigma) e^{(a+i\sigma)t} d\sigma \tag{5.40}$$

Now if we redefine (5.39) to emphasize the fact that the left hand side is really a function of $a + i\sigma$, then

$$g(a + i\sigma) = G(\sigma) = \int_{0}^{\infty} e^{-(a+i\sigma)t} f(t) dt$$

Then the equation of the transform pair is rewritten as

$$g(a + i\sigma) = \int_{0}^{\infty} e^{-(a+i\sigma)t} f(t) dt \tag{5.41}$$

where

$$f(t) = \frac{1}{2\pi} \int_{-\infty}^{\infty} e^{(a+i\sigma)t} g(a + i\sigma) d\sigma \tag{5.42}$$

Now make the following substitution $a + i\sigma = s$ such that $d\sigma = ds/i$. When $\sigma \to -\infty$, $s \to a - i\infty$; and when $\sigma \to \infty$, $s \to a + i\infty$.

Then (5.39) and (5.40) may be written as

$$g(s) = \int_{0}^{\infty} f(t) e^{-st} dt \tag{5.43}$$

$$f(t) = \frac{1}{2\pi i} \int_{a-i\infty}^{a+i\infty} g(s) e^{st} ds \tag{5.44}$$

These equations for $f(t)$ and $g(s)$ represent a Laplace transform pair. The function $g(s)$ is known as the Laplace transform of $f(t)$. The integral for $f(t)$ is known as the complex inversion integral. It is obvious that in general the parameter s which is called the Laplace parameter is a complex number. We shall continue our discussion on Laplace transform in the next chapter. Before we conclude this chapter, we will discuss in the next section the notion of convolution in the Fourier sense and then establish the relationship between Fourier and Laplace convolution.

5.2.2 Convolution integral and convolution theorem

In this section we define the convolution integral and convolution theorem, which is of considerable theoretical as well as of practical interest.

Mathematically the convolution of two functions $f(t)$ and $g(t)$, where $-\infty < t < \infty$, is defined as

$$
\begin{aligned}
y(t) &= \int_{-\infty}^{\infty} f(\tau)g(t-\tau)d\tau \\
&= \int_{-\infty}^{\infty} f(t-\tau)g(\tau)d\tau \\
&= f(t) * g(t)
\end{aligned}
\tag{5.45}
$$

This integral is usually known as the convolution integral and is denoted by $f * g$. Here τ is a dummy variable.

If $y(t)$ is the response obtained by convolving the function $f(t)$ with $g(t)$, then $\mathcal{F}\{y(t)\} = \mathcal{F}\{f * g\} = \mathcal{F}\{f(t)\}\mathcal{F}\{g(t)\}$. In words we can say that the Fourier transform of convolution of $f(t)$ and $g(t)$ is the product of the Fourier transforms $f(t)$ and $g(t)$ which is known as the convolution theorem.

5.2.3 Convolution integrals from Fourier to Laplace

The convolution integral in Laplace sense can be obtained if we suppose $f(t) = 0$ $t < 0$, and $g(t) = 0$ $t < 0$ and $f(t)$ and $g(t)$ are defined in $0 < t < \infty$. Then the convolution integral in Fourier sense can be reduced to the convolution integral in Laplace sense and is given by

$$
\begin{aligned}
y(t) &= \int_{0}^{t} f(\tau)g(t-\tau)d\tau \\
&= \int_{0}^{t} f(t-\tau)g(\tau)d\tau \\
&= f(t) * g(t)
\end{aligned}
\tag{5.46}
$$

This is called convolution or the Faltung integral and is frequently denoted by $f(t) * g(t)$. This integral has many important applications in the field of electrical engineering.

Example 6

A linear system has the impulse response given by $f(t)$ and is subjected to the rectangular pulse given by $g(t)$ as defined below.

$$
f(t) = \begin{cases} 0 & -\infty < t < 0 \\ & \qquad a > 0 \\ ae^{-at} & 0 < t < \infty \end{cases}
$$

$$
g(t) = \begin{cases} 0 & -\infty < t < -1 \\ 1 & -1 < t < 1 \\ 0 & 1 < t < \infty \end{cases}
$$

Find the output time function convolving $f(t)$ with $g(t)$ from the convolution integral

$$y(t) = \int_{-\infty}^{\infty} f(t-\tau)g(\tau)d\tau$$

using Fourier transforms.

Solution

We know

$$y(t) = \int_{-\infty}^{\infty} f(t-\tau)g(\tau)d\tau = f(t) * g(t)$$

$$\mathcal{F}\{y(t)\} = \mathcal{F}\{f(t)\}\mathcal{F}\{g(t)\}$$

Then

$$\mathcal{F}\{f(t)\} = \int_{-\infty}^{\infty} f(t)e^{-i\sigma t}dt = \int_{0}^{\infty} ae^{-at}e^{-i\sigma t}dt$$

$$\mathcal{F}\{f(t)\} = a\int_{0}^{\infty} e^{-(a+i\sigma)t}dt$$

$$= \frac{1}{a+i\sigma} = \frac{a(a-i\sigma)}{a^2+\sigma^2}$$

$$\mathcal{F}\{g(t)\} = \int_{-\infty}^{\infty} g(t)e^{-i\sigma t}dt = \int_{0}^{1} (1)e^{-i\sigma t}dt$$

$$= \frac{e^{i\sigma}-e^{-i\sigma}}{i\sigma} = \frac{2\sin\sigma}{\sigma}$$

Therefore,

$$\mathcal{F}\{y(t)\} = \mathcal{F}\{f\}\mathcal{F}\{g\}$$

$$= \frac{a(a-i\sigma)}{a^2+\sigma^2} \cdot \frac{2\sin\sigma}{\sigma}$$

$$= \frac{2a\sin\sigma}{\sigma(a^2+\sigma^2)}(a-i\sigma)$$

Taking the Fourier inverse,

$$y(t) = \frac{1}{2\pi}\int_{-\infty}^{\infty} \frac{2a\sin\sigma}{\sigma(a^2+\sigma^2)}(a-i\sigma)e^{i\sigma t}d\sigma$$

$$= \frac{a}{\pi}\int_{-\infty}^{\infty} \frac{\sin\sigma}{\sigma(a^2+\sigma^2)}(a-i\sigma)(\cos\sigma t+i\sin\sigma t)d\sigma$$

$$= \frac{a}{\pi}\int_{-\infty}^{\infty} \frac{\sin\sigma(a\cos\sigma t+\sigma\sin\sigma t)}{\sigma(a^2+\sigma^2)}d\sigma$$

$$+ \frac{ia}{\pi}\int_{-\infty}^{\infty} \frac{\sin\sigma(a\sin\sigma t-\sigma\cos\sigma t)}{\sigma(a^2+\sigma^2)}d\sigma$$

Because the integrand in the second integral is an odd function, therefore, it is zero.

Thus,

$$y(t) = \frac{a}{\pi} \int_{-\infty}^{\infty} \frac{\sin \sigma (a \cos \sigma t + \sigma \sin \sigma t)}{\sigma (a^2 + \sigma^2)} d\sigma$$

$$= \frac{2a}{\pi} \int_{0}^{\infty} \frac{\sin \sigma (a \cos \sigma t + \sigma \sin \sigma t)}{\sigma (a^2 + \sigma^2)} d\sigma$$

which is the output time function.

Exercises

1. (a) Find the Fourier transform of

$$f(t) = \begin{cases} 1/2b & |t| \le b \\ 0 & |t| > b \end{cases}$$

   (b) Determine the limit of this transform as $b \to 0$ and discuss the result.

2. Find the Fourier transform of

$$f(t) = \begin{cases} \cos \omega t & |t| \le T_0 \\ 0 & |t| > T_0 \end{cases}$$

   and sketch this transform.

3. (a) Find the Fourier transform of

$$f(t) = \begin{cases} 1 - t^2 & |t| \le 1 \\ 0 & |t| > 1 \end{cases}$$

   (b) Evaluate the integral

$$\int_{0}^{\infty} \left(\frac{t \cos t - \sin t}{t^3} \right) dt$$

4. (a) Find the Fourier sine transform of $e^{-t}, t \ge 0$.

   (b) By using the result in (a), show that

$$\int_{0}^{\infty} \frac{t \sin mt}{1 + t^2} dt = \frac{\pi}{2} e^{-m}, \quad m > 0$$

5.3 Laplace transforms

5.3.1 Introduction

In the previous section we saw that the evolution of the Laplace transform was a special case of the general Fourier transform. This section deals with the theory of Laplace transforms. We know that the method of Laplace transforms is very powerful for solving both ordinary and partial differential equations. During the 19th century, scientists and engineers were encouraged by using the operator method to solve differential equations. The operator method was first developed by the British electrical engineer Oliver Heaviside (1850–1925). In this method, operators were treated as algebraic symbols and the resulting equations were manipulated according to the rules of algebra. Remarkably, the method led to correct answers. These successes encouraged scientists and engineers to use the method even more. Some thoughtful mathematicians reasoned that there ought to be some way of placing the procedures on rigorous mathematical foundations. Research towards this goal led to the development of the method of Laplace transforms. Although Gustave Laplace, a French mathematician, initiated this method in the 18th century, Oliver Heaviside developed it and mathematicians put it in its present form.

5.3.2   Definition of Laplace transform

Let $f(t)$ be a function of t specified for $t > 0$. The Laplace transform of $f(t)$ denoted by $\mathcal{L}\{f(t)\}$ is defined by

$$\mathcal{L}\{f(t)\} \;=\; \int_0^\infty e^{-st} f(t)dt = F(s) \tag{5.47}$$

where the parameter s is, in general, a complex quantity. The real part of s is always greater than zero which guarantees the convergence of the integral provided $f(t)$ is a "well-behaved" function. The Laplace transform of $f(t)$ is usually denoted by the symbol $F(s)$.

The inverse of Laplace transforms is given by the complex inversion integral

$$f(t) = \frac{1}{2\pi i} \int_{a-i\infty}^{a+i\infty} e^{st} F(s)ds \tag{5.48}$$

where the path of integration is taken along a Bromwich contour. We shall learn about this contour when we take up the theory of residues.

For the Laplace transform of $f(t)$ defined by eqn (5.47) to exist and for its inverse, defined by (5.48), to exist, it is sufficient that the following two conditions be satisfied.

(a) The function $f(t)$ must be piecewise continuous in the interval $0 \leq t < \infty$ (Fig. 5.6). That is, every interval of the form $0 \leq t_1 \leq t \leq t_2$ can be divided into a finite number of subintervals such that $f(t)$ is continuous in the interior of each subinterval and approaches finite limits as t approaches either end-point of the interval.

(b) The function $f(t)$ must be of exponential order. That is, there is a constant λ with the property that $e^{-\lambda t}|f(t)|$ remains bounded as $t \to \infty$, i.e. there are constants λ, M and T such that

$$e^{-\lambda t}|f(t)| < M \quad \text{for} \quad t > T$$

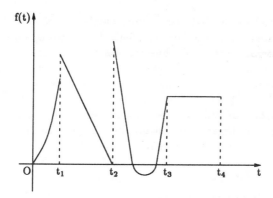

Fig. 5.6. Piecewise continuous function $f(t)$.

When these two conditions are satisfied, then a Laplace transform of $f(t)$ exists and the transform can be determined from the definition (5.48). Items (a) and (b) together define the "well-behaved" function referred to earlier.

In connection with condition (b) we need the concept of the abscissa of convergence, since for $t > 0$, $e^{-\lambda t}$ is a monotonically-decreasing function of λ. It is clear that if $e^{-\lambda t}|f(t)| < M$ $t > T$ then, for all $\lambda_1 > \lambda$, $e^{-\lambda_1 t}|f(t)| < M$ $t > T$.

Thus the λ required by condition (b) is not unique. Thus we define the abscissa of convergence of $f(t)$ as the greatest lower bound λ_0 of the set of all λ's which can be used in condition (b). For example, if $f(t) = e^{3t}$, then

$$\mathcal{L}\{f(t)\} = \int_0^\infty e^{-\lambda t} f(t) dt = \int_0^\infty e^{-(\lambda-3)t} dt$$

For the convergence of this integral λ must be greater than 3 and the greatest lower bound of λ must be $\lambda_0 = 3$. The integral may or may not exist at the greatest lower bound; in this particular example, the integral does not exist at $\lambda_0 = 3$. In a Laplace transform, we usually use s as a parameter rather than λ.

Thus we can summarize: if $f(t)$ is piecewise regular in $0 \leq t < \infty$ and of exponential order, then for any value of s which is greater than the abscissa of convergence of $f(t)$, the integral $\int_0^\infty e^{-st} f(t) dt$ converges absolutely.

5.3.3   Theorems on Laplace transforms

In the following section, we state without proof some important theorems which will be useful in obtaining the Laplace transforms.

Theorem 5.1

If $f_1(t)$ and $f_2(t)$ are piecewise regular functions of exponential order and c_1 and c_2 are any two constants, then

$$\mathcal{L}\{c_1 f_1(t) \pm c_2 f_2(t)\} = c_1 \mathcal{L}\{f_1(t)\} \pm c_2 \mathcal{L}\{f_2(t)\}$$

This result may be extended to any finite number of terms and may be extended to an infinite number if the sum is convergent. This theorem expresses the fact that the Laplace transform operator is a linear operator.

Theorem 5.2

If $f(t)$ is a continuous function of exponential order on $(0, \infty)$ whose derivative $f'(t)$ is also of exponential order and at least piecewise regular on $(0, \infty)$, then the Laplace transform of $f'(t)$ is given by the formula

$$\mathcal{L}\{f'(t)\} = s\mathcal{L}\{f(t)\} - f(0)$$

provided s is greater than the abscissa of convergence of $f(t)$.

This theorem can be extended to the nth derivative of the given function as follows. If $f(t)$ and $f'(t)$ are continuous for $0 \leq t < \infty$ and of exponential order as $t \to \infty$, while $f''(t)$ is piecewise continuous for $0 \leq t < \infty$, then

$$
\begin{aligned}
\mathcal{L}\{f''(t)\} &= \mathcal{L}\{[f'(t)]'\} \\
&= s\mathcal{L}\{f'(t)\} - f'(0); \quad \text{using theorem 5.2} \\
&= s[s\mathcal{L}\{f(t)\} - f(0)] - f'(0); \quad \text{using theorem 5.2 again}
\end{aligned}
$$

Therefore,

$$\mathcal{L}\{f''(t)\} = s^2\mathcal{L}\{f(t)\} - sf(0) - f'(0)$$

Similarly, if $f(t)$ and its first $(n-1)$ derivatives are continuous for $0 \leq t < \infty$ and of exponential order as $t \to \infty$, while $f^{(n)}(t)$ is piecewise continuous for $0 \leq t < \infty$, then

$$
\begin{aligned}
\mathcal{L}\{f^{(n)}(t)\} &= s^n\mathcal{L}\{f(t)\} - s^{n-1}f(0) - s^{n-2}f'(0) \\
&\quad - \cdots - sf^{(n-2)}(0) - f^{(n-1)}(0)
\end{aligned}
$$

In many physical situations, we find that the problem is governed by an integro differential equation. For example, the series-electrical L-C-R circuit can be described either by a differential equation or by an integrodifferential equation:

$$L\frac{dI}{dt} + IR + \frac{Q}{C} = E$$

or

$$L\frac{dI}{dt} + IR + \frac{1}{C}\int I\,dt = E$$

Theorem 5.2 can be applied directly to obtain the Laplace transform of a derivative of a function. To solve an integrodifferential equation by the Laplace transform, we need a formula. Theorem 5.3 provides the formula for solving an integrodifferential equation.

Theorem 5.3

If $f(t)$ is piecewise regular for $0 \le t < \infty$ and of exponential order, then the Laplace transform of $\int_{t_0}^{t} f(t)dt$ is given by the formula

$$\mathcal{L}\left\{\int_{t_0}^{t} f(t)dt\right\} = \frac{1}{s}\mathcal{L}\{f(t)\} + \frac{1}{s}\int_{t_0}^{0} f(t)dt \quad t \ge 0$$

The extension of this theorem to repeated integrals can be obtained as follows:

$$\mathcal{L}\left\{\int_{t_0}^{t}\int_{t_0}^{t} f(t)dtdt\right\} = \mathcal{L}\left\{\int_{t_0}^{t}\left[\int_{t_0}^{t} f(t)dt\right]dt\right\}$$

$$= \frac{1}{s}\mathcal{L}\left\{\int_{t_0}^{t} f(t)dt\right\} + \frac{1}{s}\int_{t_0}^{0}\int_{t_0}^{t} f(t)dtdt$$

$$= \frac{1}{s^2}\mathcal{L}\{f(t)\} + \frac{1}{s^2}\int_{t_0}^{0} f(t)dt + \frac{1}{s}\int_{t_0}^{0}\int_{t_0}^{t} f(t)dtdt$$

Example 1

Find the Laplace transform of the following integrodifferential equation.

$$y'(t) + ay(t) + b\int_{0}^{t} y(\lambda)d\lambda = 0; \quad y(0) = 2$$

Solution

Take the Laplace transform of the given equation:

$$\mathcal{L}\left[y' + ay + b\int_{0}^{t} y(\lambda)d\lambda\right] = \mathcal{L}(0)$$

Then, using the theorems 5.2 and 5.3,

$$\mathcal{L}\{y'\} + a\mathcal{L}\{y\} + b\mathcal{L}\left\{\int_{0}^{t} y(\lambda)d\lambda\right\} = 0$$

$$s\mathcal{L}(y) - y(0) + a\mathcal{L}\{y\} + \frac{b}{s}\mathcal{L}(y) = 0$$

Therefore $\mathcal{L}\{y\} = 2s/(s^2 + as + b)$.

5.3.4   Laplace transforms of special functions

In this section we shall give some formulae for finding the transforms of special functions. Some of the more elementary functions are the following: e^{-at}, e^{at}, $\sinh at$, $\cosh at$, $\cos bt$, $\sin bt$, t^n; and the unit step function:

$$u(t) = \begin{cases} 0 & t < 0 \\ 1 & t > 0 \end{cases}$$

The unit step function is shown in Fig. 5.7.

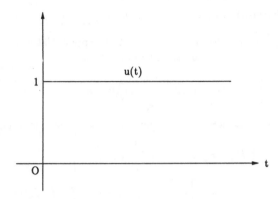

Fig. 5.7. The unit step function $u(t)$.

The transforms of these functions are the following:

Formula 1: $\mathcal{L}\{e^{-at}\} = 1/(s+a),$ $s > -a$
Formula 2: $\mathcal{L}\{e^{at}\} = 1/(s-a),$ $s > a$
Formula 3: $\mathcal{L}\{\sinh at\} = a/(s^2 - a^2),$ $s > a$
Formula 4: $\mathcal{L}\{\cosh at\} = s/(s^2 - a^2),$ $s > a$
Formula 5: $\mathcal{L}\{\cos bt\} = s/(s^2 + b^2),$ $s > 0$
Formula 6: $\mathcal{L}\{\sin bt\} = b/(s^2 + b^2),$ $s > 0$
Formula 7: $\mathcal{L}\{u(t)\} = 1/s,$ $s > 0$

Formula 8: $\mathcal{L}\{t^n\} = \begin{cases} \Gamma(n+1)/s^{n+1}, & n > -1 \\ n!/s^{n+1}, & n \text{ positive integer}, s > 0 \end{cases}$

Proofs of all these formulae except Formula 8 are left to the reader.

Proof of Formula 8

$$
\begin{aligned}
\mathcal{L}\{t^n\} &= \int_0^\infty e^{-st} t^n \, dt \\
&= \int_0^\infty e^{-z} \left(\frac{z}{s}\right)^n \frac{dz}{s}, \quad \text{put } st = z, dt = \frac{1}{s}dz \\
&= \frac{1}{s^{n+1}} \int_0^\infty e^{-z} z^n \, dz \\
&= \frac{1}{s^{n+1}} \Gamma(n+1)
\end{aligned}
$$

where $\Gamma(n+1)$ is called the **Gamma function** and is defined as

$$\Gamma(n+1) = \int_0^\infty e^{-z} z^n \, dz, \quad n+1 > 0 \tag{5.49}$$

Performing this integration by parts, we see that

$$\Gamma(n+1) = [-e^{-z}z^n]_0^\infty + n \int_0^\infty e^{-z}z^{n-1}dz$$

Since the first term on the right is zero through the end values, therefore,

$$\Gamma(n+1) = n\Gamma(n) \tag{5.50}$$

This is known as the recurrence relation of a Gamma function.
  If n is an integer, then

$$\Gamma(n+1) = (n)(n-1)(n-2)\cdots 2.1.\Gamma(1)$$

where $\Gamma(1) = \int_0^\infty e^{-z}dz = [-e^{-z}]_0^\infty = 1$ and hence $\Gamma(n+1) = n!$.
  Therefore,

$$\mathcal{L}\{t^n\} = \begin{cases} \{\Gamma(n+1)\}/s^{n+1} & n+1 > 0 \\ n!/s^{n+1} & n \text{ a positive integer} \end{cases}$$

From the definition of the Gamma function, it can be seen that

$$\Gamma(0) = \int_0^\infty e^{-z}z^{-1}dz$$

tends to ∞ at the lower limit.
  Therefore, $\Gamma(0)$ does not exist and $\Gamma(1) = 0! = 1$ by definition.
Thus

$$\begin{aligned} \Gamma(n) &= \frac{\Gamma(n+1)}{n} \\ \Gamma(-1) &= -\infty \\ \Gamma(-2) &= \infty \\ \Gamma(-3) &= -\infty \end{aligned}$$

and so on.
  It is interesting to see the numerical value of $\Gamma(\frac{1}{2})$, which is defined as

$$\Gamma\left(\frac{1}{2}\right) = \int_0^\infty e^{-z}z^{-\frac{1}{2}}dz$$

Putting $z = x^2$ so that $dz = 2xdx$, we obtain

$$\Gamma\left(\frac{1}{2}\right) = 2 \int_0^\infty e^{-x^2}dx \tag{5.51}$$

Also square of this gamma function can be written as

$$\left[\Gamma\left(\frac{1}{2}\right)\right]^2 = 4 \int_0^\infty \int_0^\infty e^{-(x^2+y^2)}dxdy$$

Substitute $x = r\cos\theta$ and $y = r\sin\theta$ and note that under this transformation, $dxdy \to rd\theta dr$.

Hence

$$\left[\Gamma\left(\frac{1}{2}\right)\right]^2 = 4\int_{r=0}^{\infty}\int_{\theta=0}^{\pi/2} e^{-r^2} rdrd\theta = \pi$$

Therefore, $\Gamma(\frac{1}{2}) = \sqrt{\pi}$.

With the knowledge of $\Gamma(\frac{1}{2})$, we are now in a position to plot the Gamma function (Fig. 5.8).

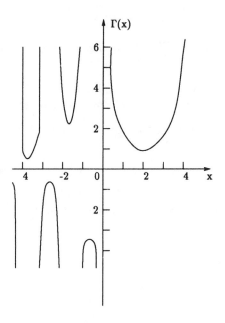

Fig. 5.8. Plot of $\Gamma(x)$ against x.

In the following, we state the three important theorems of Laplace transforms without proof. The interested reader is referred to the work of Rahman (1991).

Theorem 5.4: Scale change

If $\mathcal{L}\{f(t)\} = F(s)$ when $s > a$, then for $b > 0$,

$$\mathcal{L}\{f(bt)\} = \frac{1}{b}F\left(\frac{s}{b}\right), \quad \frac{s}{b} > a$$

Theorem 5.5: First shifting theorem

If $f(t)$ is piecewise regular for $0 \le t < \infty$ and of exponential order and if

$$\mathcal{L}\{f(t)\} = F(s) = \int_0^\infty e^{-st} f(t)dt$$

then

$$\mathcal{L}\{e^{-at} f(t)\} = F(s+a)$$

By means of this theorem the following formulas can be established:

Formula 9: $\mathcal{L}\{e^{-at}\cos bt\} = (s+a)/\{(s+a)^2 + b^2\}$
Formula 10: $\mathcal{L}\{e^{at}\cos bt\} = (s-a)/\{(s-a)^2 + b^2\}$
Formula 11: $\mathcal{L}\{e^{-at}\sin bt\} = b/\{(s+a)^2 + b^2\}$
Formula 12: $\mathcal{L}\{e^{at}\sin bt\} = b/\{(s-a)^2 + b^2\}$
Formula 13: $\mathcal{L}\{e^{-at}t^n\} = \begin{cases} \Gamma(n+1)/\{(s+a)^{n+1}\} & n+1 > 0 \\ n!/\{(s+a)^{n+1}\} & n \text{ is positive number} \end{cases}$
Formula 14: $\mathcal{L}\{e^{at}t^n\} = \begin{cases} \Gamma(n+1)/\{(s-a)^{n+1}\} & \\ n!/\{(s-a)^{n+1}\} & n \text{ is positive number} \end{cases}$

Theorem 5.6: Second shifting theorem

If $f(t)$ is piecewise continuous for $0 \le t < \infty$ and of exponential order, then

$$\mathcal{L}\{f(t-a)u(t-a)\} = e^{-as}\mathcal{L}\{f(t)\}$$

where $u(t-a)$ is a unit step function shifted from the origin by a quantity a. Note that $\mathcal{L}\{u(t-a)\} = e^{-as}/s$.

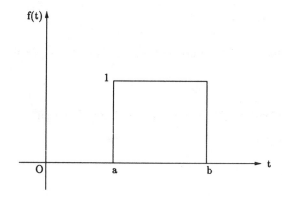

Fig. 5.9: Graph of a filter function.

Theorem 5.7: Third shifting theorem

$$\mathcal{L}\{f(t)u(t-a)\} = e^{-as}\mathcal{L}\{f(t+a)\}$$

A filter function may be mathematically defined as

$$f(t) = u(t-a) - u(t-b)$$

which can be seen in the graphical form by two unit step functions as depicted in Fig. 5.9.

Example 2

Solve the integrodifferential equation:

$$y' + 4\int_0^t y(\lambda)d\lambda = 2, \quad y(0) = 0$$

Solution

Taking the Laplace transform of the above equation,

$$\mathcal{L}\{y'\} + 4\mathcal{L}\left\{\int_0^t y(\lambda)d\lambda\right\} = \mathcal{L}\{2\}$$

$$s\mathcal{L}\{y\} - y(0) + 4\left[\frac{1}{s}\mathcal{L}\{y\} - 0\right] = \frac{2}{s}$$

Simplifying,

$$(s^2 + 4)\mathcal{L}\{y\} = 2$$

$$\mathcal{L}\{y\} = \frac{2}{s^2 + 4}$$

Therefore the inverse is given by $y(t) = \sin 2t$. This is the required solution.

Example 3

Find the Laplace transform of the following unit step function

$$f(t) = 2u(\sin \pi t) - 1$$

Solution

The graphical representation of the given function follows (Fig. 5.10).

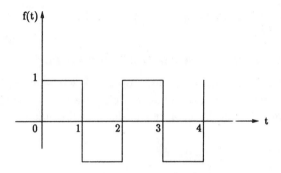

Fig. 5.10: Graphical representation of $2u(\sin \pi t) - 1$.

$$\mathcal{L}\{2u(\sin \pi t) - 1\} = \int_0^\infty e^{-st}\{2u)\sin \pi t) - 1\}dt$$

$$= \int_0^1 e^{-st}dt - \int_1^2 e^{-st}dt + \int_2^3 e^{-st}dt - \int_3^4 e^{-st}dt + \cdots$$

$$= \frac{1 - e^{-s}}{s} - \frac{e^{-s} - e^{-2s}}{s} + \frac{e^{-2s} - e^{-3s}}{s} - \frac{e^{-3s} - e^{-4s}}{s} + \cdots$$

$$= \frac{1}{s}\left[1 - 2e^{-s}\{1 - e^{-s} + e^{-2s} - e^{-3s} + \cdots\}\right]$$

$$= \frac{1}{s}\left[1 - \frac{2e^{-s}}{1 + e^{-s}}\right]$$

$$= \frac{1 + e^{-s} - 2e^{-s}}{s(1 + e^{-s})} = \frac{1 - e^{-s}}{s(1 + e^{-s})} = \frac{\{e^{s/2} - e^{-s/2}\}}{s\{e^{s/2} + e^{-s/2}\}}$$

$$= \frac{1}{s}\tanh\frac{s}{2}$$

We shall see this result again when we discuss the Laplace transform of a periodic function.

Remark

As a rule, to obtain $\mathcal{L}^{-1}\{e^{-2s}/(s+2)\}$, first find $\mathcal{L}^{-1}1/(s+2)$, which is e^{-2t}. Then, because the term e^{-2s} is a part of the transform, the t has to be changed to $(t-2)$ in e^{-2t}. The whole term must then be multiplied by the unit step function $u(t-2)$. Thus

$$\mathcal{L}^{-1}\left(\frac{e^{-2s}}{s+2}\right) = e^{-2(t-2)}u(t-2)$$

Similarly, to evaluate $\mathcal{L}^{-1}\{se^{-\pi s}/(s^2+25)\}$, we first find $\mathcal{L}^{-1}s/(s^2+25)$, which is $\mathcal{L}^{-1}\{s/(s^2+25)\} = \cos 5t$.

Then

$$\mathcal{L}^{-1}\left\{\frac{se^{-\pi s}}{s^2+25}\right\} = \cos 5(t-\pi)u(t-\pi)$$

5.3.5   Some specially important theorems

In this section we shall consider some theorems which will be useful in the application of Laplace transforms to practical problems. Before we state these theorems, we state two important properties of the Laplace transform of the function $f(t)$.

Property 1

If $f(t)$ is piecewise regular over $0 \le t < \infty$ and of exponential order, then

$$\lim_{s\to\infty} |\mathcal{L}\{f(t)\}| < \lim_{s\to\infty} \frac{M}{s-\lambda} \to 0$$

Property 2

If $f(t)$ is piecewise regular over $0 \le t < \infty$ and of exponential order, then

$$\lim_{s\to\infty} |s\mathcal{L}\{f(t)\}| < \lim_{s\to\infty} s\frac{M}{s-\lambda} = M$$

Therefore

$$\lim_{s\to\infty} |s\mathcal{L}\{f(t)\}| < M \quad \text{bounded}$$

With these two properties we state the following theorems without proof.

Theorem 5.8

If

    (i) $f(t)$ is continuous over $0 \le t < \infty$

    (ii) $\lim_{t\to 0+} f(t)$ exists

    (iii) $f'(t)$ is at least piecewise regular over $0 \le t < \infty$

    (iv) $f(t)$ and $f'(t)$ are of exponential order;

then

$$\lim_{s\to\infty} s\mathcal{L}\{f(t)\} = \lim_{t\to 0+} f(t) = f(0^+)$$

Theorem 5.9

If $f(t)$ and $f'(t)$ are piecewise regular over $0 \leq t < \infty$ and of exponential order, and $\lim_{t \to \infty} f(t)$ exists, then

$$\lim_{s \to 0} s\mathcal{L}\{f(t)\} = \lim_{t \to \infty} f(t)$$

Theorem 5.10

If $f(t)$ is piecewise regular over $0 \leq t < \infty$ and of exponential order, and if $\mathcal{L}\{f(t)\} = sG(s)$, then

$$f(t) = \frac{d}{dt}\mathcal{L}^{-1}\{G(s)\}$$

Theorem 5.11

If $f(t)$ is piecewise regular function over $0 \leq t < \infty$ and of exponential order, if $\mathcal{L}\{f(t)\} = G(s)/s$ and if $G(s)$ possesses an inverse, then

$$f(t) = \int_0^t \mathcal{L}^{-1}\{G(s)\}dt$$

Theorem 5.12

If $f(t)$ is piecewise regular over $0 \leq t < \infty$ and of exponential order and if $\mathcal{L}\{f(t)\} = F(s)$, then

$$\mathcal{L}\{tf(t)\} = -\frac{d}{ds}F(s)$$

Theorem 5.13

If $f(t)$ is piecewise regular over $0 \leq t < \infty$ and of exponential order, and if $\lim_{t \to 0}\{f(t)/t\}$ exists, and if $\mathcal{L}\{f(t)\} = F(s)$, then

$$\mathcal{L}\{\frac{f(t)}{t}\} = \int_s^\infty F(s)ds$$

Example 4:

What is

$$\mathcal{L}\left\{\int_0^t \frac{e^{-3t}\sin 2t}{t}dt\right\}?$$

Solution

$$\mathcal{L}\left\{\int_0^t \frac{e^{-3t}\sin 2t}{t}dt\right\} = \frac{1}{s}\mathcal{L}\left\{\frac{e^{-3t}\sin 2t}{t}\right\}$$

$$= \frac{1}{s}\int_s^\infty \mathcal{L}\{e^{-3t}\sin 2t\}ds$$

$$= \frac{1}{s}\int_s^\infty \frac{2}{(s+3)^2+4}ds$$

$$= \frac{1}{s}\left[\tan^{-1}\left(\frac{s+3}{2}\right)\right]_s^\infty$$

$$= \frac{1}{s}\left[\frac{\pi}{2} - \tan^{-1}\left(\frac{s+3}{2}\right)\right]$$

$$= \frac{1}{s}\cot^{-1}\left(\frac{s+3}{2}\right)$$

5.4 The unit step function and the Dirac delta function

Two generalized functions which are used frequently in applied problems are the Heaviside unit step functions, $u(t)$, and the Dirac delta function, $\delta(t)$. We have already come across the definition of the unit step function and its Laplace transform. In this section we shall see some of its properties and its relation to the Dirac delta function. It should be noted that, formerly, many writers called these two functions the unit function and the unit impulse function, respectively.

5.4.1 The unit step function and its transform

The unit step function allows us to express any function $f(t)$ which is non-zero in $a \le t \le b$ and zero elsewhere.

The equation of this function is, obviously,

$$g(t) = f(t)[u(t-a) - u(t-b)]$$

The expression $u(t-a) - u(t-b)$ is called the filter function or window function, the value of which is unity in $a \le t \le b$ and zero otherwise.

The Laplace transform of the function $g(t)$ is simply

$$\mathcal{L}\{g(t)\} = e^{-as}\mathcal{L}\{f(t+a)\} - e^{-bs}\mathcal{L}\{f(t+b)\}$$

5.4.2 The Dirac delta function

In many physical applications, it is necessary to find the response of a system to an impulsive forcing function. By an impulsive force, we mean a force that has a very large magnitude but acts over a very small period of time. An obvious example is the force encountered when a circuit is subjected to a high voltage for a short period of time, as in a lightning strike. Similarly, the force which results from the collision of two bodies can be considered to be an impulsive force. The precise distribution in time of a very short pulse, or impulse, is much less important than its total value, measured by its integral over time. Moreover, it is much easier to measure this total value than to determine the distribution in space or time. Such a distribution is called the Dirac delta function, $\delta(t)$, after the British theoretical physicist P.A.M. Dirac (1902–).

It is defined by

$$\delta(t-h) = \begin{cases} 0 & t \neq h \\ \infty & t = h \end{cases} \tag{5.52}$$

$$\int_{-\infty}^{\infty} \delta(t-h)dt = 1 \tag{5.53}$$

Readers who wish to know more about its properties should refer to Dirac, *The Principles of Quantum Mechanics*, Oxford University Press (1935).

We will arrive at a definition of an impulse function by starting with the definition of a rectangular pulse of height $1/h$ and width h, as shown in Fig. 5.11.

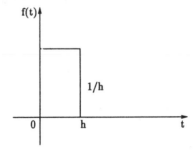

Fig. 5.11: A rectangular pulse.

$$\delta_h(t) = \frac{1}{h}[u(t) - u(t-h)], \quad h > 0 \tag{5.54}$$

where $u(t)$ and $u(t-h)$ are two unit step functions. Thus $\delta_h(t)$ can be expressed as

$$\delta_h(t) = \begin{cases} 1/h & 0 \leq t \leq h \\ 0 & t > h \end{cases} \tag{5.55}$$

Now in the limiting process, when $t \to 0$, eqn (5.54) reduces to

$$\delta(t) = \lim_{h \to 0} \delta_h(t) = \begin{cases} \infty & t = 0 \\ 0 & t \neq 0 \end{cases} \tag{5.56}$$

The physical implication of (5.54) can be visualized from Fig. 5.11. We allow the pulse to shrink in width and, at the same time, to increase in height, so that its area remains constant. Continuing this process, we shall eventually obtain an extremely narrow, very large amplitude pulse at $t = 0$. If we proceed to the limit, where the width approaches zero and the height approaches infinity (but still with the product width X height $= 1$), we approach the delta function $\delta(t)$ defined in (5.55).

Thus, the strength of the $\delta(t)$ function is given by

$$\int_0^\infty \delta(t)dt = 1 \qquad\qquad (5.57)$$

This may be written as $\int_0^\infty \delta(t-h)dt = 1$ if the source of the impulse is at $t = h$. An important property of the impulse function is expressed as

$$\int_a^b f(t)\delta(t-h)dt = f(h), \quad a < h < b \qquad\qquad (5.58)$$

Thus, the Dirac delta function can be related to the Heaviside step function as follows:

$$\delta(t) \;=\; \lim_{h\to 0}\frac{u(t)-u(t-h)}{h} = \frac{d}{dt}u(t) \qquad\qquad (5.59)$$

The unit impulse is only the first of infinitely many singular functions. As a generalization of the unit impulse, we can define the unit doublet, triplet, etc., and we will not pursue this any further.

Example 1

Solve the initial value problem

$$y' + 2y \;=\; 3u(t-1) - \delta(t-2)$$
$$y(0) \;=\; 1$$

Solution

Taking the Laplace transform of the given equation,

$$\mathcal{L}\{y'\} + 2\mathcal{L}\{y\} \;=\; 3\mathcal{L}\{u(t-1)\} - \mathcal{L}\{\delta(t-2)\}$$

$$s\mathcal{L}\{y\} - y(0) + 2\mathcal{L}\{y\} \;=\; 3\frac{e^{-s}}{s} - e^{-2s}$$

$$(s+2)\mathcal{L}\{y\} \;=\; 1 + 3\frac{e^{-s}}{s} - e^{-2s}$$

$$\mathcal{L}\{y\} \;=\; \frac{1}{s+2} + 3\frac{e^{-s}}{s(s+2)} - \frac{e^{-2s}}{s+2}$$

Therefore,

$$y = e^{-2t} + \frac{3}{2}\{1 - e^{-2(t-1)}\}u(t-1) - e^{-2(t-2)}u(t-2)$$

This is the required solution.

5.4.3 Properties of generalized functions

We have already seen the graphical behaviour of a Dirac delta function. A smooth curve can be drawn, as shown in Fig. 5.12.

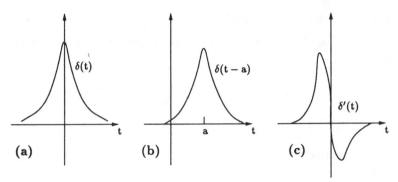

Fig. 5.12: Generalized functions:
(a) $\delta(t)$; (b) $\delta(t-a)$; (c) $\delta'(t)$.

Under appropriate conditions, an ordinary function $f(t)$ can also be multiplied by a generalized function. Thus, we have the following properties:

$$\delta(t)f(t) = f(t)\delta(t) = f(0)\delta(t)$$
$$\delta(t-a)f(t) = f(t)\delta(t-a) = f(a)\delta(t-a)$$

provided $f(t)$ is any continuous function at $t = 0$ and $t = 1$ respectively.

(I) Differentiation rule of generalized functions

Consider

$$y = f(t)u(t)$$

Then

$$\begin{aligned}
y' &= f'(t)u(t) + f(t)u'(t) \\
&= f'(t)u(t) + f(t)\delta(t) \\
&= f'(t)u(t) + f(0)\delta(t)
\end{aligned}$$

An expression for $f(t)\delta'(t)$ is deduced from the product rule as follows. We know

$$[f(t)\delta(t)]' = [f(0)\delta(t)]' = f(0)\delta'(t)$$

Also

$$\begin{aligned}
[f(t)\delta(t)]' &= f'(t)\delta(t) + f(t)\delta'(t) \\
&= f'(0)\delta(t) + f(t)\delta'(t)
\end{aligned}$$

Equating these two expressions gives

$$f(0)\delta'(t) = f'(0)\delta(t) + f(t)\delta'(t)$$

or,

$$f(t)\delta'(t) = f(0)\delta'(t) - f'(0)\delta(t)$$

provided f and f' are both continuous at $t = 0$. In the same way, we may deduce the rules:

$$f(t)\delta'(t-a) = f(a)\delta'(t-a) - f'(a)\delta(t-a)$$
$$f(t)\delta''(t-a) = f(a)\delta''(t-a) - 2f'(a)\delta'(t-a) + f''(a)\delta(t-a)$$

(II) Integration rule of generalized functions

Generalized functions can also be integrated. For instance, if $a < b$ and $a \neq 0$ and $b \neq 0$, then

1. $\int_a^b \delta(t)dt = \int_a^b u'(t)dt = u(b) - u(a) = \begin{cases} 1, & a < 0 < b \\ 0, & \text{otherwise} \end{cases}$

2. $\int_a^b \delta'(t)dt = \delta(b) - \delta(a) = 0$

3. $\int_a^b \delta''(t)dt = 0$

4. $\int_a^b g(t)\delta(t)dt = g(0)\int_a^b \delta(t)dt = g(0), \quad a < 0 < b$

5. $\int_a^b g(t)\delta'(t)dt = \int_a^b [g(0)\delta'(t) - g'(0)\delta(t)]dt = -g'(0), \quad a < 0 < b$

6. $\int_a^b g(t)\delta^{(k)}(t-c)dt = (-1)^k g^k(c), \quad a < c < b$ provided $g^k(c)$ is continuous at $t = c$.

7. $\delta(-t) = \delta(t)$ is an even function.

8. $\delta'(-t) = -\delta'(t)$ is an odd function.

Example 2

If the charge $Q(t)$ is given by $Q(t) = \sin\sqrt{LC}tu(t)$, then find the current, $I(t)$.

Solution

$$Q(t) = \sin\sqrt{LC}tu(t)$$

Then

$$\begin{aligned} I(t) &= \frac{dQ}{dt} \\ &= \sqrt{LC}\cos\sqrt{LC}tu(t) + \sin\sqrt{LC}t\delta(t) \\ &= \sqrt{LC}\cos\sqrt{LC}tu(t) \end{aligned}$$

Exercises

1. The sine and cosine transforms of $f(t)$ are defined respectively as

$$S\{f(t)\} = \int_0^\pi f(t) \sin ntdt, \quad n = 1, 2, 3, \cdots$$

and

$$C\{f(t)\} = \int_0^\pi f(t) \cos ntdt, \quad n = 0, 1, 2, \cdots$$

Obtain formulae expressing $S\{f'\}$ and $C\{f'\}$ in terms of $S\{f\}$ and $C\{f\}$.

2. Consider that $T\{f(t)\}$ is a general integral transform

$$T\{f(t)\} = \int_a^b f(t) K(s,t) dt$$

where $K(s,t)$ is the so-called kernel of the transformation. Obtain the conditions for $K(s,t)$ so that $T\{f'\}$ and $T\{f''\}$ contain no terms involving the evaluation of f or any derivatives of f. Find at least one kernel satisfying these conditions.

3. Determine the response of the damped vibrating system corresponding to

$$y'' + 3y' + 2y = f(t), \quad y(0) = 0, \quad y'(0) = 0$$

where

$$f(t) = \begin{cases} 1 & 0 < t < 1 \\ 0 & t > 1 \end{cases}$$

4. An LCR circuit has a switch replacing the current $I(t)$. When $t < 0$, the switch is open and C has a charge Q_0. Show that the damped oscillatory current $I(t)$ at any time after closing the switch is $I(t) = -(Q_0/LC\omega_n)e^{-\frac{Rt}{2L}}\sin \omega_n t$, where the frequency $\omega_n = \sqrt{(1/LC) - (R^2/4L^2)}$. (Hint: Use the Laplace transform method.)

5. Solve each of the following differential equations so that the solutions satisfy the given initial conditions:

   - $ty'' - ty' - y = 0, \quad y_0 = 0, \ y_0' = 1$
   - $y'' + 2ty' - 4y = 1, \quad y_0 = y_0' = 1$
   - $ty'' + 2(2t - 1)y' - 4y = 0, \quad y_0 = 1, \ y_0' = 0$

6. Find the current $I(t)$ in an LC circuit where $L = 1$ henry, $C = 1$ farad, zero initial current and charge on the capacitor, and $E(t)$ as follows:

(a) $E(t) = \begin{cases} t & 0 < t < a \\ 0 & t > a \end{cases}$

(b) $E(t) = \begin{cases} t & 0 < t < a \\ a & t > a \end{cases}$

(c) $E(t) = \begin{cases} 1 - e^{-t} & 0 < t < \pi \\ 0 & t > \pi \end{cases}$

7. Find $f'(t)$ as a generalized function from the given function $f(t)$ and then verify that $\mathcal{L}(f'(t)) = s\mathcal{L}(f(t))$:

   (a) $(2t + 5)u(t)$
   (b) $\sin 2t u(t)$
   (c) $u(t) - u(t - 2)$
   (d) $u(t) - 5\delta(t)$
   (e) $2u(t) - \delta(t) + \delta'(t - 1)$

8. Find the inverse transform of the following functions:

   (a) $a + s$
   (b) $3 + 2s + s^2$
   (c) $(s^2 + a^2)/(s^2 + b^2)$
   (d) $e^{-2s}(3 + 4s)$
   (e) $e^{-s}\{(s + 3)/(s - 5)\}$

9. An inductor of 1 henry and a capacitor of 0.01 farad are in series with an AC generator of $10 \cos 10t$. At $t = 0$ the charge on the capacitor and the current in the circuit are zero. Show that there is an electrical **resonance** in the circuit and that the charge at any time is given by $(t/2) \sin 10t$. [Hint: Recall that when the applied frequency coincides with the natural frequency, the phenomenon of resonance occurs.]

5.5   The Heaviside expansion theorems for finding inverses

The partial fraction plays a very important role in obtaining inverse Laplace transforms. In the following section, we extend this technique and systematically write this procedure in the form of those theorems usually known as the Heaviside expansion theorems, after the British electrical engineer Oliver Heaviside. We state the theorems without proof. The interested reader is referred to the standard textbooks, including Rahman (1991).

Theorem 5.14

If

$$\mathcal{L}\{f(t)\} = \frac{P(s)}{Q(s)}$$

where $P(s)$ and $Q(s)$ are polynomials and the degree of $Q(s)$ is greater than the degree of $P(s)$, and $Q(s)$ has n distinct zeros $a_1, a_2, \cdots, a_n$, then

$$f(t) = \mathcal{L}^{-1}\left\{\frac{P(s)}{Q(s)}\right\} = \sum_{k=1}^{n} \frac{P(a_k)}{Q'(a_k)} e^{a_k t}$$

Theorem 5.15

If

$$\mathcal{L}\{f(t)\} = \frac{P(s)}{Q(s)}$$

where $P(s)$ and $Q(s)$ are polynomials, and where $Q(s) = 0$ has a repeated root b of multiplicity r while the remaining roots, $a_1, a_2, \cdots, a_n$, do not repeat, then

$$
\begin{aligned}
f(t) &= \mathcal{L}^{-1}\left\{\frac{P(s)}{Q(s)}\right\} \\
&= \left[A_1 + A_2 t + \frac{A_3 t^2}{2!} + \cdots + A_r \frac{t^{r-1}}{(r-1)!}\right] e^{bt} + \sum_{k=1}^{n} \frac{P(a_k)}{Q'(a_k)} e^{a_k t}
\end{aligned}
$$

where

$$A_{r-k} = \lim_{s \to b} \frac{1}{k!} \frac{d^k}{ds^k}\left[(s-b)^r \frac{P(s)}{Q(s)}\right] \quad k = 0, 1, 2, \cdots, (r-1)$$

Theorem 5.16

If

$$f(t) = \mathcal{L}^{-1}\left\{\frac{P(s)}{Q(s)}\right\}$$

where $P(s)$ and $Q(s)$ are polynomials and the degree of $Q(s)$ is greater than the degree of $P(s)$, then the terms in $f(t)$ which correspond to an unrepeated complex factor $(s+a)^2 + b^2$ of $Q(s)$ are

$$\frac{e^{-at}}{b}(\phi_i \cos bt + \phi_r \sin bt)$$

where ϕ_r and ϕ_i are, respectively, the real and imaginary parts of $\phi(-a+ib)$ and where $\phi(s)$ is the quotient of $P(s)$ and all the factors of $Q(s)$ except the factor $(s+a)^2 + b^2$.

Theorem 5.17

If

$$f(t) = \mathcal{L}^{-1}\left\{\frac{P(s)}{Q(s)}\right\}$$

where $P(s)$ and $Q(s)$ are polynomials and the degree of $Q(s)$ is greater than the degree of $P(s)$, then the terms in $f(t)$ which correspond to a repeated complex factor $[(s+a)^2 + b^2]^2$ of $Q(s)$ are

$$\frac{e^{-at}}{b}\{Cb\cos bt + (D - aC)\sin bt\} + \frac{Ae^{-at}}{2b}t\sin bt$$

$$+ \frac{B - aA}{2b^3}e^{-at}[\sin bt - bt\cos bt]$$

where

$$A = \frac{\phi_i}{b}$$

$$B = \frac{1}{b}[b\phi_r + a\phi_i]$$

$$C = \frac{1}{2b^3}[\phi_i - b\phi'_r]$$

$$D = \frac{1}{2b^3}[b^2\phi'_i - ab\phi'_r + a\phi_i]$$

in which ϕ_r and ϕ_i are, respectively, the real and imaginary parts of $\phi(-a + ib)$ and where $\phi(s)$ is the quotient of $P(s)$ and all the factors of $Q(s)$ except the factor $[(s+a)^2 + b^2]^2$; a prime denotes the differentiation.

Example 1

Find

$$\mathcal{L}^{-1}\left\{\frac{s}{(s+2)^2(s^2 + 2s + 10)}\right\}$$

Solution

Here $p(s) = s$ and $q(s) = (s+2)^2(s^2 + 2s + 10) = (s+2)^2\{(s+1)^2 + 3^2\}$.

Thus $q(s) = 0$ yields $s = -2$; two repeated roots and $s = -1 \pm i3$; two complex roots. Now the inverse corresponding to $s = -2$ is $(A_1 + A_2t)e^{-2t}$ where

$$A_2 = \lim_{s\to-2}\left\{(s+2)^2\frac{s}{(s+2)^2(s^2 + 2s + 10)}\right\} = -\frac{1}{5}$$

and

$$A_1 = \lim_{s \to -2} \frac{d}{ds} \left\{ (s+2)^2 \frac{s}{(s+2)^2(s^2+2s+10)} \right\}$$

$$= \lim_{s \to -2} \left[\frac{1}{(s^2+2s+10)} - \frac{2s+2}{(s^2+2s+10)^2} \right]$$

$$= \lim_{s \to -2} \frac{-s^2+10}{(s^2+2s+10)^2} = \frac{3}{50}$$

Thus the inverse corresponding to $s = -2$ is $\{(3/50) - (t/5)\}e^{-2t}$.

To obtain the inverse corresponding to the complex roots $s = -1 \pm i3$ $(a = 1, b = 3)$, we know (according to theorem 5.16) that $\phi(s) = s/(s+2)^2$.

Thus $\phi(-1+i3) = (13 - 9i)/50$ and equating real and imaginary parts we obtain $\phi_r = 13/50$ and $\phi_i = -9/50$.

Thus the inverse corresponding to the factor $(s^2 + 2s + 10)$ is

$$\frac{e^{-t}}{3} \left\{ \frac{-9\cos 3t + 13\sin 3t}{50} \right\} = \frac{e^{-t}}{150} \{-9\cos 3t + 13\sin 3t\}$$

Hence,

$$\mathcal{L}^{-1} \left\{ \frac{s}{(s+2)^2(s^2+2s+10)} \right\} = \left(\frac{3}{50} - \frac{1}{5}t \right) e^{-2t} + \frac{e^{-t}}{150}(-9\cos 3t + 13\sin 3t)$$

5.5.1 Laplace transform of a periodic function

In this section we shall state and prove a theorem concerning the application of Laplace transforms to general periodic functions.

Theorem 5.18

If $f(t)$ is of exponential order and is a periodic function with period $2p$ on $(0, \infty)$ and is piecewise regular over $0 \le t \le 2p$, then

$$\mathcal{L}\{f(t)\} = \frac{\int_0^{2p} f(t)e^{-st} dt}{1 - e^{-2ps}} \quad s > 0$$

Example 2

Find the Laplace transform of the function shown in Fig. 5.13.

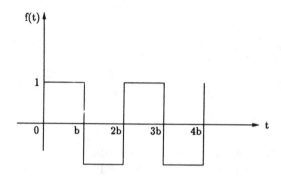

Fig. 5.13: An alternating rectangular wave.

Solution

The period of the given function is $2b$.

$$\mathcal{L}\{f(t)\} = \frac{1}{1 - e^{-2bs}} \int_0^{2b} f(t)e^{-st}\,dt$$

$$= \frac{1}{1 - e^{-2bs}} \left[\int_0^b e^{-st}\,dt - \int_b^{2b} e^{-st}\,dt \right]$$

$$= \frac{1}{1 - e^{-2bs}} \left[\left(-\frac{e^{-st}}{s}\right)_0^b - \left(-\frac{e^{-st}}{s}\right)_b^{2b} \right]$$

$$= \frac{1}{1 - e^{-2bs}} \left[\frac{1 - 2e^{-bs} + e^{-2bs}}{s} \right]$$

$$= \frac{(1 - e^{-bs})^2}{s(1 + e^{-bs})(1 - e^{-bs})} = \frac{1 - e^{-bs}}{s(1 + e^{-bs})}$$

$$= \frac{e^{bs/2} - e^{-bs/2}}{s(e^{bs/2} + e^{-bs/2})}$$

$$= \frac{1}{s} \tanh\left(\frac{bs}{2}\right)$$

Example 3

Applying the convolution theorem, solve the following initial value problem,

$$y'' + y = \sin t \qquad y(0) = 0, \quad y'(0) = 0$$

Solution

Taking the Laplace transform of the given equation, we have

$$\mathcal{L}(y'') + \mathcal{L}(y) = \mathcal{L}(\sin t)$$

$$s^2 \mathcal{L}\{y\} - sy_0 - y_0' + \mathcal{L}\{y\} = \frac{1}{s^2 + 1}$$

$$(s^2 + 1)\mathcal{L}\{y\} = \frac{1}{s^2 + 1}$$

Therefore

$$\mathcal{L}\{y\} = \frac{1}{(s^2 + 1)^2}$$

Then

$$
\begin{aligned}
y &= \mathcal{L}^{-1}\left\{\frac{1}{(s^2 + 1)^2}\right\} \\
&= \int_0^t \sin \tau \sin(t - \tau)d\tau \\
&= \frac{1}{2}\int_0^t [\cos(t - 2\tau) - \cos t]dt \\
&= \frac{1}{2}[\sin t - t \cos t]
\end{aligned}
$$

This is the required solution.

Exercises

Find the Laplace inverse of each of the following transforms by the Heaviside theorems or otherwise.

1. $1/\{(s + 1)(s + 2)\}$

2. $1/\{(s + 1)(s - 2)\}$

3. $s/(s^2 + a^2)$

4. $2s/(s^2 - 3s + 2)$

5. $21/\{(s - 1)(s^2 + 4s + 9)\}$

6. $(3s - 1)/\{(s^2 - 2s + 5)^2\}$

7. $(4s^2 + 2)/\{s(s - 1)(s^2 + 4)\}$

8. $1/(s^2 + 4) + 5/\{(s^2 + 4)^2\}$

9. $(3s + 2)/[(s + 1)^2 + 1]^2$

10. Find the solutions of each of the following differential equations where $f(t)$ is in each case a periodic function defined over one period as indicated:

- $y' + 2y + \int_0^t y\, dt = f(t), \quad y_0 = 1$ where $f(t)$ is defined below.
- $y'' + 4y = f(t)$, where

$$f(t) = \begin{cases} 1 & 0 < t < \pi/2 \\ 0 & \pi/2 < t < \pi \end{cases}$$

$y_0 = y_0' = 0$

5.6 Transfer function, impulse response and indicial response

The application of the convolution theorem makes it possible to determine the response of a system to a general excitation. To develop this idea we need the concepts of:

(a) transfer function (system function)

(b) impulsive response

(c) indicial response (indicial admittance).

Any physical system capable of responding to an excitation $f(t)$ can be described mathematically by a differential equation satisfying the output function $y(t)$. Assume that all initial conditions are zero when a single excitation $f(t)$ begins to act. Then, taking the Laplace transform of the governing differential equation, we obtain a relationship in the form,

$$\mathcal{L}\{y(t)\} = \frac{1}{Z(s)}\mathcal{L}\{f(t)\} \tag{5.60}$$

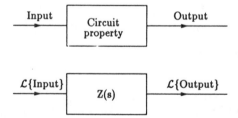

Fig. 5.14: Transfer function.

In the electrical network problem $f(t)$ is the input voltage, and $y(t)$ is the output current known as the response. Here $1/Z(s)$ is defined as the transfer function of the system. In pictorial notation, this transfer function can be defined as Fig. 5.14, i.e.,

$$\frac{1}{Z(s)} = \frac{\mathcal{L}\{y(t)\}}{\mathcal{L}\{f(t)\}} = \frac{\mathcal{L}\{\text{output}\}}{\mathcal{L}\{\text{input}\}}$$

The importance of $Z(s)$ functions is apparent in systems other than the electrical system.

The output of a system to a unit impulse input is called the impulsive response of the system, and is usually denoted by $I(t)$. Importantly, if $I(t)$ is known, then the response to any general input may be obtained from the convolution integral.

One immediate result can be deduced as follows. We know that if $f(t)$ is the general excitation, and $y(t)$ is the response, then

$$\mathcal{L}\{y(t)\} = \frac{1}{Z(s)}\mathcal{L}\{f(t)\}$$

Let $f(t) = \delta(t) =$ unit impulse, the response of $\delta(t) = I(t)$; hence,

$$\mathcal{L}\{I(t)\} = \frac{1}{Z(s)}\mathcal{L}\{\delta(t)\} = \frac{1}{Z(s)} \qquad (5.61)$$

The transfer function is equal to the transform of the impulsive response.

It is more convenient, from a practical point of view, to deal with the response of a unit step function $u(t)$. This is called the indicial response and is normally denoted by $h(t)$. If the indicial response is known, then the response to a general input can be determined from the Duhamel theorem. This theorem will be discussed later.

In the relationship (5.60), let $f(t) = u(t) =$ unit step function. Then the response of $u(t) = h(t)$, and, hence,

$$\mathcal{L}\{h(t)\} = \frac{1}{Z(s)}\mathcal{L}\{u(t)\} = \frac{1}{sZ(s)} \qquad (5.62)$$

The significance of the impulsive response and the indicial response is that once these two responses are determined (easily accomplished through measuring), then the responses of the system can be determined by the convolution integral and the Duhamel theorem respectively.

5.6.1 Significance of convolution theorem

Given an impulsive response $I(t)$ corresponding to a unit impulse, we obtain from (5.61)

$$\mathcal{L}\{y(t)\} = \frac{1}{Z(s)}\mathcal{L}\{f(t)\} = \mathcal{L}\{I(t)\}\mathcal{L}\{f(t)\} \qquad (5.63)$$

Then by the convolution integral, we have

$$y(t) = \int_0^t I(\tau)f(t-\tau)d\tau = \int_0^t I(t-\tau)f(\tau)d\tau$$

$$\text{output} = \text{(Input)} * \text{(Impulsive Response)} \qquad (5.64)$$

5.6.2   Significance of Duhamel's theorem

An alternate method of determining the output response of a system is provided by Duhamel's theorem, named after the French mathematician (1797–1872), and used when the indicial response is known. The response to unit step function is normally easier to determine than that to unit impulse functions.

Now rearranging (5.60) with the help of (5.62) we obtain

$$
\begin{aligned}
\mathcal{L}\{y(t)\} &= s\mathcal{L}\{h(t)\}\mathcal{L}\{f(t)\} \\
&= \mathcal{L}\{h(t)\}\left[\mathcal{L}\left\{\left(\frac{df}{dt}\right)+f(0)\right\}\right] \\
&= \mathcal{L}\{h(t)\}\mathcal{L}\left\{\left(\frac{df}{dt}\right)\right\}+f(0)\mathcal{L}\{h(t)\}
\end{aligned}
\tag{5.65}
$$

or,

$$
\begin{aligned}
\mathcal{L}\{y(t)\} &= \left[\mathcal{L}\left\{\left(\frac{dh}{dt}\right)+h(0)\right\}\right]\mathcal{L}\{f(t)\} \\
&= \mathcal{L}\left\{\left(\frac{dh}{dt}\right)\right\}\mathcal{L}\{f(t)\}+h(0)\mathcal{L}\{f(t)\}
\end{aligned}
\tag{5.66}
$$

Then applying the convolution theorem to (5.65) and (5.66), we obtain

$$
\begin{aligned}
y(t) &= f(0)h(t)+\int_0^t \frac{df}{dt}(\tau)h(t-\tau)d\tau
\end{aligned}
\tag{5.67}
$$

$$
\begin{aligned}
&= f(t)h(0)+\int_0^t \frac{dh}{dt}(\tau)f(t-\tau)d\tau
\end{aligned}
\tag{5.68}
$$

Equations (5.67) and (5.68) express the output in terms of the input and the indicial response. These formulas are normally called Duhamel's Theorem.

We shall illustrate the usefulness of the convolution theorem and Duhamel's theorem by a practical example of network theory.

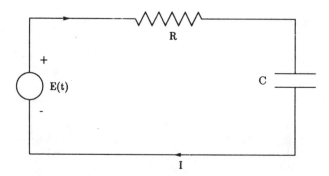

Fig. 5.15: *RC* circuit.

Example 1

Assuming that the RC circuit shown in Fig. 5.15 is dead initially, find the current which flows when the voltage applied is $\sin \omega t u(t)$,

(a) by finding the system function and using the Laplace transform;

(b) by using the impulsive response method; and

(c) by using the indicial response method.

Solution

The differential equation governing the system can be written as

$$Ry + \frac{1}{C}\int_0^t y\,dt = f(t), y(0) = 0 \tag{5.69}$$

where $y(t)$ is the output current and $f(t)$ is the input voltage. Then, taking the Laplace transform of the given equation, we obtain

$$R\mathcal{L}\{y\} + \frac{1}{C}\mathcal{L}\left\{\int_0^t y\,dt\right\} = \mathcal{L}\{\sin \omega t\}$$

$$R\mathcal{L}\{y\} + \frac{1}{Cs}\mathcal{L}\{y\} = \frac{\omega}{s^2 + \omega^2}$$

or,

$$(RCs + 1)\mathcal{L}\{y\} = \frac{\omega Cs}{s^2 + \omega^2}$$

Therefore,

$$\mathcal{L}\{y\} = \left(\frac{Cs}{1 + Rcs}\right)\left(\frac{\omega}{s^2 + \omega^2}\right)$$

Here the transform function $= Cs/(1 + RCs)$

(a) By transfer function method:

$$y = \mathcal{L}^{-1}\left\{\frac{(\omega/R)s}{(s + \frac{1}{RC})(s^2 + \omega^2)}\right\}$$

The poles are at $s = -1/RC$ and $s = \pm i\omega$. The inverse is obtained by the theory of residue.

$$R_1 = \text{residue at } s = -\frac{1}{RC}$$

$$= \lim_{s \to -1/RC}\left\{\left(s + \frac{1}{RC}\right)\frac{(\omega/R)se^{st}}{(s + 1/RC)(s^2 + \omega^2)}\right\}$$

$$= -\frac{\omega C e^{-t/RC}}{1 + (RC\omega)^2}$$

$$R_2 = \text{residue at } s = +i\omega$$

$$= \lim_{s \to i\omega}\left\{(s - i\omega)\frac{(\omega/R)se^{st}}{(s + 1/RC)(s + i\omega)(s - i\omega)}\right\}$$

$$= \frac{\omega}{R}\frac{i\omega e^{i\omega t}}{(i\omega + 1/RC)2i\omega} = \frac{\omega}{2R}\frac{e^{i\omega t}}{(i\omega + 1/RC)}$$

Therefore

$$R_2 = \frac{\omega(1/RC - i\omega)e^{+i\omega t}}{2R(\omega^2 + 1/R^2C^2)}$$

$$R_3 = \text{residue at } s = -i\omega$$

$$= \frac{\omega(1/RC + i\omega)e^{-i\omega t}}{2R(\omega^2 + 1/R^2C^2)}$$

$$R_2 + R_3 = \frac{\omega}{2R(\omega^2 + 1/R^2C^2)}\left[\frac{1}{RC}(e^{i\omega t} + e^{-i\omega t}) - i\omega(e^{i\omega t} - e^{-i\omega t})\right]$$

$$= \frac{\omega}{2R(\omega^2 + 1/R^2C^2)}\left[\frac{2}{RC}\cos\omega t + 2\omega\sin\omega t\right]$$

Thus the solution is obtained as

$$y(t) = \frac{-\omega C e^{-t/RC}}{1 + (RC\omega)^2} + \frac{\omega C}{[1 + (RC\omega)^2]}[\cos\omega t + \omega RC\sin\omega t] \qquad (5.70)$$

This is the required response.

(b) By impulsive response method:

$$\mathcal{L}\{y\} = \left(\frac{\frac{s}{R}}{\frac{1}{RC} + s}\right)\left(\frac{\omega}{s^2 + \omega^2}\right)$$

We know

$$\mathcal{L}\{I(t)\} = \left(\frac{\frac{s}{R}}{s + \frac{1}{RC}}\right) = \frac{1}{R}\left[1 - \frac{1}{RC}\frac{1}{(s + \frac{1}{RC})}\right]$$

Then $I(t) = (1/R)\left[\delta(t) - (1/RC)e^{-t/RC}\right]$.

Now using the convolution integral

$$y(t) = \int_0^t I(\tau)f(t - \tau)d\tau$$

$$= \int_0^t \frac{1}{R}\left\{\delta(\tau) - \frac{1}{RC}e^{-(\tau/RC)}\right\}\sin\omega(t - \tau)d\tau$$

$$= \frac{1}{R}\int_0^t \delta(\tau)\sin\omega(t - \tau)d\tau - \frac{1}{R^2C}\int_0^t e^{-(\tau/RC)}\sin\omega(t - \tau)d\tau$$

$$\text{The first integral} \quad = \quad \frac{1}{R}\int_0^t \delta(\tau)\sin\omega(t-\tau)d\tau$$

$$= \quad \frac{1}{R}\sin\omega t$$

$$\text{The second integral} \quad = \quad \frac{1}{R^2C}\int_0^t e^{-\frac{1}{RC}(t-\tau)}\sin\omega\tau d\tau$$

$$= \quad \frac{1}{R^2C}e^{-t/RC}\int_0^t e^{\tau/RC}\sin\omega\tau d\tau$$

$$= \quad \frac{e^{-t/RC}}{R^2C\,(\omega^2+1/(R^2C^2))} \times$$

$$\left[e^{t/RC}\left\{\frac{1}{RC}\sin\omega t - \omega\cos\omega t\right\}+\omega\right]$$

Therefore,

$$-\frac{1}{R^2C}\int_0^t e^{-\tau/RC}\sin\omega(t-\tau)d\tau$$

$$= \quad \frac{-\omega e^{-t/RC}}{R^2C\,(\omega^2+1/(R^2C^2))}+\frac{[-\sin\omega t+\omega RC\cos\omega t]}{R(1+\omega^2 R^2 C^2)}$$

$$= \quad \frac{-\omega e^{-t/RC}}{1+\omega^2 R^2 C^2}+\frac{(-\sin\omega t+\omega RC\cos\omega t)}{R(1+\omega^2 R^2 C^2)}$$

Therefore,

$$y(t) \quad = \quad \frac{-\omega C e^{-t/RC}}{1+\omega^2 R^2 C^2}+\frac{-\sin\omega t+\omega RC\cos\omega t+(1+\omega^2 R^2 C^2)\sin\omega t}{R(1+\omega^2 C^2 R^2)}$$

$$= \quad \frac{-\omega C e^{-t/RC}}{1+\omega^2 R^2 C^2}+\frac{\omega C[\cos\omega t+\omega CR\sin\omega t]}{1+\omega^2 C^2 R^2} \tag{5.71}$$

(c) **By indicial response method:**

We know that

$$\mathcal{L}\{h(t)\} \quad = \quad \left(\frac{\frac{s}{R}}{s+\frac{1}{RC}}\right)\mathcal{L}\{u(t)\}=\frac{(s/R)}{)}s+(1/RC)\frac{1}{s}=\frac{(1/R)}{s+(1/RC)}$$

Therefore $h(t)=(1/R)e^{-t/RC}$.

Then, using Duhamel's formula,

$$y(t) \quad = \quad f(0)h(t)+\int_0^t \frac{df}{d\tau}h(t-\tau)d\tau$$

$$= \quad \int_0^t \omega\cos\omega\tau\frac{1}{R}e^{-(1/RC)(t-\tau)}d\tau$$

$$= \quad \frac{\omega}{R}e^{-t/RC}\int_0^t e^{\tau/RC}\cos\omega\tau d\tau$$

$$= \frac{\omega}{R} e^{-t/RC} \left[\frac{e^{\tau/RC} \{ (1/RC) \cos \omega\tau + \omega \sin \omega\tau \}}{\omega^2 + (1/RC)^2} \right]_0^t$$

$$= \frac{\frac{-\omega}{R} e^{-t/RC}}{\omega^2 + (1/RC)^2} \left[e^{t/RC} \left(\frac{1}{RC} \cos \omega t + \omega \sin \omega t \right) + \frac{1}{RC} \right]_0^t$$

$$y(t) = \frac{-\omega C e^{-t/RC}}{1 + \omega^2 R^2 C^2} + \frac{\omega C [\cos \omega t + \omega RC \sin \omega t]}{1 + \omega^2 R^2 C^2} \tag{5.72}$$

It can be noted from (5.69), (5.70) and (5.71) that the responses $y(t)$ produced by $\sin \omega t$ are identical. In this particular problem, Duhamel's formula yields the result very easily, while the other methods seem laborious.

Exercises

Using the convolution theorem, find the inverse of each of the following transforms:

1. $1/(s^2 + a^2)^2$

2. $s/(s^2 + 4)^3$

3. $(s^2 + 2s + 1)/(s^2 + 2s + 5)^2$

4. $1/(s\sqrt{s-1})$

5. $1/\{\sqrt{s}(s^2 - 4)\}$

6. $1/\{s^2(s+a)\}$

7. $(1/s) = (1/\sqrt{s}).(1/\sqrt{s})$

8. Find the current which flows in a series RC circuit when the voltage applied is $\cos \omega t u(t)$

   - by finding the transfer function, and using Laplace transforms, and
   - by finding the initial admittance and using the convolution integral.

5.7   Complex inversion integrals

In this section, the theory of residues will be used to obtain the inverse Laplace transform. For convenience we rewrite the Laplace transform pair defined in (5.47) and (5.48).

$$\mathcal{L}(f(t)) = \int_0^\infty e^{-st} f(t) dt \tag{5.73}$$

and by the complex inversion integral, we showed that:

$$f(t) = \frac{1}{2\pi i} \int_{a-i\infty}^{a+i\infty} \mathcal{L}(f(t)) e^{st} ds \tag{5.74}$$

where s is a complex variable.

Formula (5.73) is usually known as the complex inversion integral. It is also known as Bromwich's integral formula. This result provides a direct means for obtaining the inverse Laplace transform when its transform is known.

The integral in (5.73) is to be performed along a line $Re(s) = a$ in the complex plane $s = x + iy$.

The real number a is arbitrary and is chosen so that $s = a$ lies to the right of all singularities such as poles, branch points, essential singular points, etc.

The integral is evaluated by considering the closed contour known as Bromwich contour:

$$\frac{1}{2\pi i} \oint_C e^{st} \mathcal{L}(f(t)) ds \tag{5.75}$$

where C is the contour shown in Fig. 5.16. This closed contour is composed of the line AB and the arc BDA of a circle of radius R with centre at the origin. In the following we state and prove the necessary theorem to demonstrate the method of residues for finding inverse Laplace transforms.

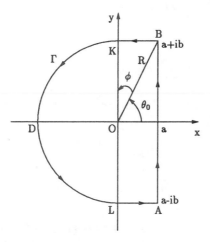

Fig. 5.16: A Bromwich contour.

Theorem 5.19

If the Laplace transform $F(s)$ is an analytic function of s, except at a finite number of singular points called the poles (each of which lies to the left of the vertical line $Re(s) = a$), and if $sF(s)$ is bounded as s approaches infinity through the half plane $Re(s) \leq a$, then

$$\mathcal{L}^{-1}\{F(s)\} = \frac{1}{2\pi i} \int_{a-i\infty}^{a+i\infty} F(s) e^{st} ds$$

$$= \sum \text{residues of } F(s)e^{st} \text{ at all the poles of } F(s)$$
$$= R_1 + R_2 + \cdots + R_n$$

where R_k's are defined as residues and are given by the formula

$$R_k = \lim_{s \to a_k} \{(s - a_k)F(s)e^{st}\} \qquad k = 1, 2, \cdots, n$$

if $a_1, a_2, a_3, \cdots, a_n$ are simple poles of $F(s)$. Now if $s = a_k$ is a multiple pole of $F(s)$ of order $(r + 1)$, then

$$R_k = \lim_{s \to a_k} \frac{1}{r!} \frac{d^r}{ds^r} \{(s - a_k)^{r+1} F(s)e^{st}\}$$

Proof

Using the Bromwich contour shown in Fig. 5.16, we have by residue theorem:

$$\frac{1}{2\pi i} \int_{a-ib}^{a+ib} F(s)e^{st}ds + \frac{1}{2\pi i} \int_\Gamma F(s)e^{st}ds = \sum \text{residues of } F(s)e^{st} \qquad (5.76)$$

Hence we can write:

$$\left| \frac{1}{2\pi i} \int_{a-ib}^{a+ib} F(s)e^{st}ds - \sum \text{residues of } F(s) \; e^{st} \right| = \left| \frac{-1}{2\pi i} \int_\Gamma F(s)e^{st}ds \right| \qquad (5.77)$$

Now, if Γ_1, Γ_2, Γ_3, and Γ_4 represent arcs BK, KD, DL and LA respectively, then:

$$\int_\Gamma F(s)e^{st}ds = \int_{\Gamma_1} F(s)e^{st}ds + \int_{\Gamma_2} F(s)e^{st}ds + \int_{\Gamma_3} F(s)e^{st}ds +$$
$$\int_{\Gamma_4} F(s)e^{st}ds \qquad (5.78)$$

Now, if we can prove that each of these integrals on the right approaches zero as $R \to \infty$, we will have proved the result. To do this, we consider the four integrals. Now along Γ_1 we have:

$$s = Re^{i\theta} \qquad \theta_0 \le \theta \le \frac{\pi}{2}$$
$$I_1 = \int_{\Gamma_1} e^{st}F(s)ds$$
$$= \int_{\theta_0}^{\pi/2} e^{Re^{i\theta}t} \; F(Re^{i\theta})iR \; e^{i\theta} d\theta$$

$$|I_1| \leq \int_{\theta_0}^{\pi/2} |e^{(R\cos\theta)t}||e^{i(R\sin\theta)t}||F(Re^{i\theta})||iR\ e^{i\theta}|\ d\theta$$

$$\leq \int_{\theta_0}^{\pi/2} e^{(R\cos\theta)t}|F(Re^{i\theta})|\ Rd\theta$$

Now suppose that on Γ we have

$$|F(s)| < \frac{M}{R^k}$$

where $k > 0$ and M and R are constants on Γ. Then:

$$|I_1| \leq \frac{M}{R^{k-1}} \int_{\theta_0}^{\pi/2} e^{(R\cos\theta)t} d\theta$$

But,

$$\theta = \frac{\pi}{2} - \phi$$

$$|I_1| \leq \frac{M}{R^{k-1}} \int_0^{\phi_0} e^{(R\sin\phi)t} d\phi$$

where $\phi_0 = (\pi/2) - \theta_0 = \sin^{-1}(a/R)$. Since $\phi < \phi_0 = \pi/2$ which is evident from Fig. 5.17, and consequently we have $\sin\phi \leq \sin\phi_0 \leq \cos\theta_0 = a/R$, then this integral is less than or equal to:

$$\frac{M}{R^{k-1}} \int_0^{\phi_0} e^{at} d\phi = \frac{Me^{at}}{R^{k-1}} \phi_0 = \frac{Me^{at}}{R^{k-1}} \sin^{-1}\left(\frac{a}{R}\right)$$

But, as $R \to \infty$, $\sin^{-1}(a/R) \to (a/R)$, thus $\lim_{R\to\infty} |I_1| = 0$.

Along Γ_2 we have:

$$I_2 = \int_{\Gamma_2} e^{st} F(s)ds \qquad s = Re^{i\theta},\ \frac{\pi}{2} \leq \theta \leq \pi$$

$$= \int_{\pi/2}^{\pi} e^{(Re^{i\theta}t)}\ F(R\ e^{i\theta})i\ R\ e^{i\theta} d\theta$$

Thus we have:

$$|I_2| \leq \frac{M}{R^{k-1}} \int_{\pi/2}^{\pi} e^{(R\cos\theta)t} d\theta$$

$$\leq \frac{M}{R^{k-1}} \int_0^{\pi/2} e^{-(R\sin\phi)t} d\phi$$

where $\theta = (\pi/2) + \phi$.

Note that the graph of the function $y = \sin\phi$ is depicted in Fig. 5.17 which shows the arc of the sine curve and the line OA, i.e. $y = 2\phi/\pi$. It is geometrically evident that $\sin\phi > 2\phi/\pi$ for the range $0 \leq \phi \leq \pi/2$.

But, since $\sin\phi \geq 2\phi/\pi$ for $0 \leq \phi \leq \pi/2$, therefore

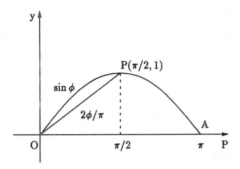

Fig. 5.17: Graph of $\sin\phi$ and $\phi/(\pi/2)$.

$$|I_2| \leq \frac{M}{R^{k-1}} \int_0^{\pi/2} e^{-(2R\phi t)/\pi} d\phi$$

$$|I_2| \leq \frac{\pi M}{2tR^k}(1 - e^{-Rt})$$

Thus,

$$\lim_{R\to\infty} I_2 = 0$$

From symmetry, we can conclude that integrals along Γ_3 and Γ_4 become zero. Thus when $R \to \infty$, and consequently $b \to \infty$, we can write:

$$f(t) = \frac{1}{2\pi i} \int_{a-i\infty}^{a+i\infty} F(s)e^{st} ds$$

$$= \sum \text{residues of } F(s)e^{st}$$

Hence this is the required proof.

The validity of Theorem 5.19 can be ascertained on the assumption that the integral around Γ approaches zero as $R \to \infty$. A sufficient condition under which this assumption is correct is that:

$$|f(s)| < \frac{M}{R^k}$$

where $M > 0$ and $R > 0$. Then

$$\lim_{R\to\infty} \int_\Gamma e^{st} F(s) ds = 0$$

This condition always holds if $F(s) = P(s)/Q(s)$ where $P(s)$ and $Q(s)$ are polynomials and the degree of $P(s)$ is less than the degree of $Q(s)$.

Example 1

Find

$$\mathcal{L}^{-1}\left\{\frac{s^2 - s + 3}{s^3 + 6s^2 + 11s + 6}\right\}$$

Solution

It can easily be shown that

$$\lim_{s\to\infty} sF(s) = \lim_{s\to\infty}\left\{\frac{s(s^2 - s + 3)}{s^3 + 6s^2 + 11s + 6}\right\}$$
$$= 1$$

which is bounded. Therefore, by Theorem 7.19, we know

$$\mathcal{L}^{-1}\left\{\frac{s^2 - s + 3}{s^3 + 6s^2 + 11s + 6}\right\} = \mathcal{L}^{-1}\{F(s)\}$$
$$= \sum \text{residues of } F(s)e^{st}$$
$$\text{at poles } s = -1, -2, \text{and } -3$$

These are the simple poles.

$$R_1 = \text{residue at } s = -1$$
$$= \lim_{s\to-1}\left\{(s+1)\frac{(s^2 - s + 3)e^{st}}{(s+1)(s+2)(s+3)}\right\} = \frac{5}{2}e^{-t}$$
$$R_2 = \text{residue at } s = -2$$
$$= \lim_{s\to-2}\left\{(s+2)\frac{(s^2 - s + 3)e^{st}}{(s+1)(s+2)(s+3)}\right\} = -9e^{-2t}$$
$$R_3 = \text{Residue at } s = -3$$
$$= \lim_{s\to-3}\left\{(s+3)\frac{(s^2 - s + 3)e^{st}}{(s+1)(s+2)(s+3)}\right\} = \frac{15}{2}e^{-3t}$$

Then

$$\mathcal{L}^{-1}\left\{\frac{s^2 - s + 3}{(s+1)(s+2)(s+3)}\right\} = R_1 + R_2 + R_3$$
$$= \frac{5}{2}e^{-t} - 9e^{-2t} + \frac{15}{2}e^{-3t}$$

Example 2

Evaluate $\mathcal{L}^{-1}\{1/\{(s+1)(s-2)^2\}\}$ by using the method of residues.

Solution

We can write

$$\mathcal{L}^{-1}\frac{1}{(s+1)(s-2)^2} = \frac{1}{2\pi i}\int_{a-i\infty}^{a+i\infty}\frac{e^{st}\ ds}{(s+1)(s+2)^2}$$

$$= \sum \text{residue of } \frac{e^{st}}{(s+1)(s-2)^2}$$

at the poles $s = -1$ and $s = 2$.

Now

$$R_1 = \text{residue at simple pole } s = -1$$

$$= \lim_{s\to -1}\frac{e^{st}}{(s-2)^2} = \frac{1}{9}e^{-t}$$

$$R_2 = \text{residue at double pole } s = 2$$

$$= \lim_{s\to 2}\frac{d}{ds}\left\{\frac{e^{st}}{s+1}\right\}$$

$$= \lim_{s\to 2}\left\{\frac{te^{st}}{s+1} - \frac{e^{st}}{(s+1)^2}\right\}$$

$$= \frac{1}{3}te^{2t} - \frac{e^{2t}}{9}$$

$$\mathcal{L}^{-1}\left\{\frac{1}{(s+1)(s-2)^2}\right\} = \frac{1}{9}e^{-t} + \left(\frac{1}{3}t - \frac{1}{9}\right)e^{2t}$$

Example 3

Evaluate $\mathcal{L}^{-1}(e^{-k\sqrt{s}})/s$ by the use of the complex inversion formula.

Solution

By the complex inversion formula, the required Laplace inverse can be written

$$f(t) = \frac{1}{2\pi i}\int_{a-i\infty}^{a+i\infty}\frac{e^{-k\sqrt{s}}}{s}e^{st}ds \qquad (5.79)$$

$$= \frac{1}{2\pi i}\int_{a-i\infty}^{a+i\infty}F(s)ds$$

Since $s = 0$ is a branch point of the integral, we have to modify the Bromwich contour and the modified contour is shown in Fig. 5.18.

By Cauchy's theorem:

$$\frac{1}{2\pi i}\left[\int_{AB} F(s)ds + \int_{BC} F(s)ds + \int_{CD} F(s)ds\right.$$

$$\left. + \int_{DEF} F(s)ds + \int_{FG} F(s)ds + \int_{GA} F(s)ds\right] = 0 \qquad (5.80)$$

BC and GA are the arcs of a circle of radius R and centre at 0, and the arc DEF of a circle of radius ε with centre at 0.

Now according to theorem 5.19 when $R \to \infty$ $\int_{BC} F(s)ds \to 0$ and $\int_{GA} F(s)ds \to 0$.

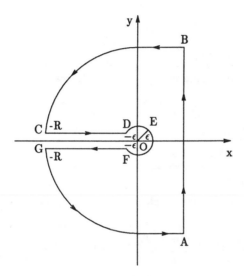

Fig. 5.18: A Bromwich contour which encloses a branch point.

Therefore,

$$f(t) = \lim_{R\to\infty,\epsilon\to 0} \frac{1}{2\pi i} \int_{AB} F(s)ds = \frac{1}{2\pi i} \int_{a-i\infty}^{a+i\infty} F(s)ds$$

$$= -\lim_{R\to\infty,\ \epsilon\to 0} \frac{1}{2\pi i}\left\{\int_{CD} F(s)ds + \int_{DEF} F(s)ds\right.$$

$$\left. + \int_{FG} F(s)ds\right\} \qquad (5.81)$$

In the following we demonstrate the procedure to evaluate these three integrals. We strictly follow the arrow in the process of integration.

Now along CD,

$$s = xe^{\pi i} = -x$$
$$\sqrt{s} = \sqrt{x}e^{\pi i/2} = i\sqrt{x}$$

and as s goes from $-R$ to $-\varepsilon$, x goes from R to ε. Thus, we have:

$$\int_{CD} \frac{e^{st-\sqrt{s}k}}{s}ds = \int_{-R}^{-\varepsilon} \frac{e^{st-k\sqrt{s}}}{s}ds$$
$$= \int_{R}^{\varepsilon} \frac{e^{-xt-ki\sqrt{x}}}{-x}(-dx)$$

Similarly, along FG,

$$s = xe^{-\pi i} = -x$$
$$\sqrt{s} = \sqrt{x}e^{-\pi i/2} = -i\sqrt{x}$$

and as s goes $-\varepsilon$ to $-R$, x goes from ε to R and hence,

$$\int_{FG} \frac{e^{st-k\sqrt{s}}}{s}ds = \int_{-\varepsilon}^{-R} \frac{e^{st-k\sqrt{s}}}{s}ds$$
$$= \int_{\varepsilon}^{R} \frac{e^{-xt+ki\sqrt{x}}}{x}dx$$

Also, along DEF, $s = \varepsilon e^{i\theta}$ and we have

$$\int_{DEF} \frac{e^{st-k\sqrt{s}}}{s}ds = \int_{\pi}^{-\pi} \frac{e^{\varepsilon e^{i\theta}t-k\sqrt{\varepsilon}e^{i\theta/2}}}{\varepsilon e^{i\theta}}\varepsilon i e^{i\theta}d\theta$$
$$= i\int_{\pi}^{-\pi} e^{\varepsilon e^{i\theta}t-k\sqrt{\varepsilon}e^{i\theta/2}}d\theta$$

Thus, gathering all this information, eqn (5.81) becomes:

$$f(t) = \frac{1}{2\pi i}\int_{a-i\infty}^{a+i\infty} \frac{e^{st-k\sqrt{s}}}{s}ds$$
$$= -\lim_{R\to\infty,\,\varepsilon\to 0} \frac{1}{2\pi i}\left\{\int_{R}^{\varepsilon} \frac{e^{-xt-ik\sqrt{x}}}{x}dx + \int_{\varepsilon}^{R} \frac{e^{-xt+ik\sqrt{x}}}{x}dx\right.$$
$$\left. +i\int_{\pi}^{-\pi} e^{\varepsilon e^{i\theta}t-k\sqrt{\varepsilon}e^{i\theta/2}}d\theta\right\}$$
$$= -\lim_{R\to\infty,\,\varepsilon\to 0} \frac{1}{2\pi i}\left\{\int_{\varepsilon}^{R} \frac{e^{-xt}(e^{ik\sqrt{x}} - e^{-ik\sqrt{x}})}{x}dx\right.$$
$$\left. +i\int_{\pi}^{-\pi} e^{\varepsilon e^{i\theta}t-k\sqrt{\varepsilon}e^{i\theta/2}}d\theta\right\}$$

$$= -\lim_{R\to\infty,\, \varepsilon\to 0} \frac{1}{2\pi i}\left\{2i\int_\varepsilon^R \frac{e^{-xt}\{\,sin k\sqrt{x}\}}{x}dx\right.$$

$$\left. +i\int_\pi^{-\pi} e^{\varepsilon e^{i\theta}t - k\sqrt{\varepsilon}e^{i\theta/2}}\,d\vartheta\right\}$$

Now

$$\lim_{\varepsilon\to 0}\int_\pi^{-\pi} e^{\varepsilon e^{i\theta}t - k\sqrt{\varepsilon}e^{i\theta/2}}\,d\theta = -2\pi$$

Thus we find:

$$f(t) \;=\; 1 - \frac{1}{\pi}\int_0^\infty \frac{e^{-xt}\{sin\ k\sqrt{x}\}}{x}dx$$

Using the result presented in example 4, this can be written as:

$$f(t) = 1 - erf\left(\frac{k}{2\sqrt{t}}\right) = erfc\left(\frac{k}{2\sqrt{t}}\right)$$

where $erf(y) = (2/\sqrt{\pi})\int_0^y e^{-\eta^2}\,d\eta$.

Example 4

Show that $(1/\pi)\int_0^\infty\{(e^{-xt}sin\ k\sqrt{x})/x\}dx = erf\left(k/(2\sqrt{t})\right)$

Solution

Let us denote that $I = (1/\pi)\int_0^\infty\{(e^{-xt}sin\ k\sqrt{x})/x\}dx$.

Substituting $x = u^2$ such that $dx = 2udu$, and then differentiating the resulting integral with respect to k two times, we obtain

$$\frac{\partial I}{\partial k} = \frac{2}{\pi}\int_0^\infty e^{-u^2 t}\cos kudu$$

$$\frac{\partial^2 I}{\partial k^2} = -\frac{k}{2t}\frac{\partial I}{\partial k}$$

It is easily observed that the initial conditions can be obtained as $I = 0$ and $(\partial I/\partial k) = (1/\sqrt{\pi t})$ when $k = 0$. Integrating the above partial differential equation with respect to k using these initial conditions we obtain the required result.

Example 5

Find $\mathcal{L}^{-1}\left\{\frac{1}{\sinh bs}\right\}$.

Solution

The poles are given by the roots of the equation $\sinh bs = 0$ which are simply $s = (\pm n\pi i)/b$ for $n = 0, 1, 2, \cdots$.

The residue at $s = 0$ is

$$R_0 = \lim_{s \to 0} \frac{se^{st}}{\sinh bs} = \frac{1}{b}$$

The residue at $s = (n\pi i)/b$ (using L'Hospital's rule) is

$$R_n = \lim_{s \to \frac{n\pi i}{b}} \left(s - \frac{n\pi i}{b}\right) \frac{e^{st}}{\sinh bs}$$

$$= \frac{e^{n\pi it/b}}{b\cosh n\pi i}$$

$$= (-1)^n \frac{e^{\frac{n\pi it}{b}}}{b}$$

In a similar manner the residue at $s = -n\pi i/b$ is simply $R_{-n} = (-1)^n \frac{e^{-n\pi it/b}}{b}$.

Hence the required inverse is given by

$$f(t) = \mathcal{L}^{-1}\left\{\frac{1}{\sinh bs}\right\}$$

$$= \frac{1}{b} + \frac{1}{b}\sum_{n=1}^{\infty}[(-1)^n(e^{n\pi it/b} + e^{-n\pi it/b})]$$

$$= \frac{1}{b} + \frac{2}{b}\sum_{n=1}^{\infty}(-1)^n \cos\frac{n\pi t}{b}$$

This is the required solution.

Exercises

Using the method of residues, find the inverses of the following Laplace transforms:

1. $\mathcal{L}^{-1}\left\{\dfrac{1}{(s+a)^2 + b^2}\right\}$

2. $\mathcal{L}^{-1}\left\{\dfrac{1}{(s+2)(s+3)}\right\}$

3. $\mathcal{L}^{-1}\left\{\dfrac{1}{(s+3)^2}\right\}$

4. $\mathcal{L}^{-1}\left\{\dfrac{1}{s^2+9}\right\}$

5. $\mathcal{L}^{-1}\left\{\dfrac{1}{s^2+4s+13}\right\}$

6. $\mathcal{L}^{-1}\left\{\dfrac{1}{s^2(s^2+1)}\right\}$

7. $\mathcal{L}^{-1}\left\{\dfrac{s}{s^3+1}\right\}$

8. $\mathcal{L}^{-1}\left\{\dfrac{1}{(s^2+9)(s^2+16)}\right\}$

9. $\mathcal{L}^{-1}\left\{\dfrac{1}{(s^2+2s+5)^2}\right\}$

Find the inverse of each of the following Laplace transforms:

10. $\mathcal{L}^{-1}\left\{\dfrac{1}{s\cosh bs}\right\}$

11. $\mathcal{L}^{-1}\left\{\dfrac{1}{s\sinh bs}\right\}$

12. $\mathcal{L}^{-1}\left\{\dfrac{1}{(s+a)\sinh bs)}\right\}$

13. $\mathcal{L}^{-1}\left\{\dfrac{1}{(s+a)\cosh bs)}\right\}$

14. $\mathcal{L}^{-1}\left\{\dfrac{\sinh x\sqrt{s}}{s\sinh\sqrt{s}}\right\}$

5.8   Stability criteria

In the analysis of many practical problems we should not need to know the complete behaviour of the solution system. It is sufficient for such system to know whether the solution is stable, i.e. whether its response corresponding to a finite excitation remains bounded or becomes infinite as t goes to infinity. This question can be answered very successfully by analysing the Laplace transform of the response without actually determining the response itself.

Let us then consider the Laplace transform of the response of the system as

$$\mathcal{L}\{y(t)\} \equiv F(s)$$

where $F(s)$ is a rational function, i.e.

$$F(s) = \frac{P(s)}{Q(s)}$$

in which P and Q are real polynomials in the complex variable $s = a + iw$. We know from the Heaviside theorem that the inverse transform

$$y(t) = \mathcal{L}^{-1}\left\{\frac{P(s)}{Q(s)}\right\}$$

is determined completely and solely by the factors of $Q(s)$. It is clear that the roots of the polynomial equation $Q(s) = 0$ play a very important role in the stability of the response function. As for example, if $\mathcal{L}(y(t)) = a/(s^2 + a^2)$ then $y(t) = \mathcal{L}^{-1}\{a/(s^2 + a^2)\} = \sin at$. This response $y(t)$ is stable for any t. However, if $\mathcal{L}y(t) = a/(s^2 - a^2)$, then

$$y(t) = \mathcal{L}^{-1}\left\{\frac{a}{s^2 - a^2}\right\} = \sinh at = \frac{1}{2}(e^{at} - e^{-at})$$

which is not stable when $t \to \infty$ for any a. But if $\mathcal{L}y(t) = 1(s + a)$ then $y(t) = \mathcal{L}^{-1}\{1/(s + a)\} = e^{-at}$. This function is stable for $a > 0$ when $t \to \infty$. Likewise there are many transforms whose inverses do not produce stable solution.

We can list the following transforms whose inverses are easily determined.

A. The unrepeated roots of $Q(s)$ in the transforms and their inverses are:

1. $\mathcal{L}^{-1}\left\{\dfrac{1}{s}\right\} = 1, u(t)$

2. $\mathcal{L}^{-1}\left\{\dfrac{\omega}{s^2 + \omega^2}\right\} = \sin \omega t$

3. $\mathcal{L}^{-1}\left\{\dfrac{s}{s^2 + \omega^2}\right\} = \cos \omega t$

4. $\mathcal{L}^{-1}\left\{\dfrac{1}{s - a}\right\} = e^{at}$

5. $\mathcal{L}^{-1}\left\{\dfrac{\omega}{(s - a)^2 + \omega^2}\right\} = e^{at}\sin \omega t$

6. $\mathcal{L}^{-1}\left\{\dfrac{s - a}{(s - a)^2 + \omega^2}\right\} = e^{at}\cos \omega t$

B. The repeated roots of $Q(s)$ in the transforms and their inverses are:

7. $\mathcal{L}^{-1}\left\{\dfrac{1}{s^{n+1}}\right\} = \dfrac{t^n}{\Gamma(n + 1)}, n + 1 > 0$

8. $\mathcal{L}^{-1}\left\{\dfrac{1}{(s^2 + \omega^2)^2}\right\} = \dfrac{1}{2\omega^3}(\sin \omega t - \omega t \cos \omega t)$

9. $\mathcal{L}^{-1}\left\{\dfrac{s}{(s^2 + \omega^2)^2}\right\} = \dfrac{t}{2\omega}\sin \omega t$

Thus the forms of $\mathcal{L}^{-1}1/((s^2 + \omega^2)^n)$ may be written as $t^n \cos \omega t$ or $t^n \sin \omega t$ for $n > 1$.

10. $\mathcal{L}^{-1}\left\{\dfrac{1}{(s-a)^{n+1}}\right\} = \dfrac{e^{at}t^n}{\Gamma(n+1)}, n+1>0$

11. $\mathcal{L}^{-1}\left\{\dfrac{\omega^2}{((s-a)^2+\omega^2)^2}\right\} = \dfrac{e^{at}}{2\omega}(\sin\omega t - \omega t\cos\omega t)$

12. $\mathcal{L}^{-1}\left\{\dfrac{(s-a)^2}{((s-a)^2+\omega^2)^2}\right\} = \dfrac{e^{at}}{2\omega}(\sin\omega t + \omega t\cos\omega t)$

Thus the forms of $\mathcal{L}^{-1}\left\{1/((s-a)^2+\omega^2)^n)\right\}$ may be written as $e^{at}t^n\cos\omega t$, $e^{at}t^n\sin\omega t$ for $n>1$.

In the above analysis, it is evidently clear that the Laplace transforms of the forms 1, 2 and 3 are always **stable** although they do not approach zero when t goes to infinity. The Laplace inverses of the forms 4, 5 and 6 are **stable** if and only if $a<0$. They not only remain finite but in fact approach to zero when t goes to infinity. Terms of the forms 7, 8 and 9 are **unstable** in all circumstances, because of the multiplying factor t as t goes to infinity, and in fact the response becomes unbounded. The terms of the forms 10, 11 and 12 are **stable** if and only if $a<0$.

Hence translating these observations into conditions on the roots of the polynomial equation $Q(s)=0$, we see that the response $y(t)$ of a system will be **stable** if and only if the following conditions are satisfied:

(I) Every unrepeated real root is nonpositive.

(II) Every repeated real root is negative.

(III) Every pure imaginary root is unrepeated.

(IV) Every general complex root has a negative real part.

These conditions can be described by the following theorem:

Theorem 5.20

In order for the response function

$$y(t) = \mathcal{L}^{-1}\left\{\dfrac{P(s)}{Q(s)}\right\}$$

to be stable where t goes to infinity it is necessary and sufficient that the polynomial equation $Q(s)=0$ have no roots to the right of the imaginary axis in the complex s plane and that any roots on the imaginary axis in the s plane be unrepeated.

To study the behaviour of the roots of a polynomial equation, the reader is referred to the works of Bronwell (1953) and Guilleman (1953).

A preliminary result of considerable importance is contained in the following theorem:

Theorem 5.21

If the real part of each root of the polynomial equation $Q(s) = 0$ is less than or equal to zero, then the coefficients in $Q(s)$ will have the same sign.

However, if the coefficients in $Q(s)$ all have the same sign, the real part of each root of $Q(s) = 0$ may or may not be less than or equal to zero.

As for example, the roots of the following polynomial equation $Q(s) = s^4 + s^3 + s^2 + 11s + 10 = 0$, the coefficients of which all have the same sign, are the following

$$s = -1, -2, 1 \pm 2i$$

The two complex roots have the positive real parts.

For cubic equations, a sufficient condition, supplementing Theorem 4.5, is contained in the following result:

Theorem 5.22

A necessary and sufficient condition that every root of the cubic equation $a_0 s^3 + a_1 s^2 + a_2 s + a_3 = 0$ have negative real parts is that all coefficients have the same sign and that

$$a_1 a_2 - a_0 a_3 > 0$$

The proof is left to the reader as an exercise.

The extension of Theorem 5.22 to polynomial equations of higher degree is contained in the following theorem.

Theorem 5.23

In the polynomial equation

$$Q(s) = a_0 s^n + a_1 s^{n-1} + \cdots + a_{n-1} s + a_n = 0$$

let every coefficient be positive, and construct the n determinants such that

$$D_1 = a_1, D_2 = \begin{vmatrix} a_1 & a_0 \\ a_3 & a_2 \end{vmatrix}, D_3 = \begin{vmatrix} a_1 & a_0 & 0 \\ a_3 & a_2 & a_1 \\ a_5 & a_4 & a_3 \end{vmatrix}, \cdots$$

$$D_n = \begin{vmatrix} a_1 & a_0 & 0 & 0 & 0 & 0 & \cdots & 0 & 0 \\ a_3 & a_2 & a_1 & a_0 & 0 & 0 & \cdots & 0 & 0 \\ \cdots & \cdots & \cdots & \cdots & \cdots & \cdots & \cdots & \cdots & \cdots \\ a_{2n-1} & a_{2n-2} & a_{2n-3} & \cdots & \cdots & \cdots & \cdots & \cdots & a_n \end{vmatrix}$$

where, in each determinant, all a's with negative subscripts or subscripts greater than n are to be replaced by zero.

Then a necessary and sufficient condition that each root of $Q(s) = 0$ have negative real part is that each D_i be positive. The test provided by Theorem 5.23 is commonly known as the Routh stability criterion or Routh and Hurwitz stability criterion.

Example 1

Determine whether the following equation has any roots with negative real parts:

$$Q(s) = s^4 + s^3 + s^2 + 10s + 10$$

Solution

In this problem we have, $a_0 = 1, a_1 = 1, a_2 = 1, a_3 = 10$ and $a_4 = 10$.

Now we form the determinants as follows:

$$D_1 = 1, D_2 = \begin{vmatrix} 1 & 1 \\ 10 & 1 \end{vmatrix}, D_3 = \begin{vmatrix} 1 & 1 & 0 \\ 10 & 1 & 1 \\ 0 & 10 & 10 \end{vmatrix}, D_4 = \begin{vmatrix} 1 & 1 & 0 & 0 \\ 10 & 1 & 1 & 1 \\ 0 & 10 & 10 & 1 \\ 0 & 0 & 0 & 10 \end{vmatrix}$$

Evaluating these determinants, we obtain $D_1 = 1, D_2 = -9, D_3 = -100$ and $D_4 = -1000$. Since all Ds are not positive, the given equation has at least one root whose real part is non-negative. Hence, $\mathcal{L}^{-1}\{P(s)/Q(s)\}$ has the response which is not **stable**.

In the following theorem, we shall discuss a somewhat different method of obtaining information about the location of roots of an equation $f(z) = 0$.

Theorem 5.24

If $f(z)$ is analytic within and on a closed contour C except at a finite number of poles, and if $f(z)$ has neither poles nor zeros on C, then

$$\frac{1}{2\pi i} \oint_C \frac{f'(z)}{f(z)} dz = N - P$$

where N is the number of zeros of $f(z)$ within C, and P is the number of poles of $f(z)$ within C, each counted as many times as its multiplicity.

Proof

Suppose first that at a point $z = a_k$ within C, $f(z)$ has a zero of order n_k which itself is a positive integer.

Then $f(z)$ can be written as

$$f(z) = (z - a_k)^{n_k} \phi(z)$$

where $\phi(z)$ is analytic in some neighbourhood of $z = a_k$.

Thus we have

$$f'(z) = n_k(z - a_k)^{n_k-1} \phi(z) + (z - a_k)^{n_k} \phi'(z)$$

Therefore

$$\frac{f'(z)}{f(z)} = \frac{n_k(z-a_k)^{n_k-1}\phi(z) + (z-a_k)^{n_k}\phi'(z)}{(z-a_k)^{n_k}\phi(z)}$$

$$= \frac{n_k}{z-a_k} + \frac{\phi'(z)}{\phi(z)}$$

Here $\phi'(z)/\phi(z)$ is analytic at $z = a_k$. Similarly if $f(z)$ has a pole of order p_k which itself is a positive integer at $z = b_k$, we can write

$$f(z) = (z-b_k)^{-p_k}\psi(z)$$

where $\psi(z)$ is analytic in some neighbourhood of $z = b_k$.

We obtain

$$f'(z) = -p_k(z-b_k)^{-p_k-1}\psi(z) + (z-b_k)^{-p_k}\psi'(z)$$

Therefore

$$\frac{f'(z)}{f(z)} = \frac{-p_k(z-b_k)^{-p_k-1}\psi(z) + (z-b_k)^{-p_k}\psi'(z)}{(z-b_k)^{-p_k}\psi(z)}$$

$$= \frac{-p_k}{z-b_k} + \frac{\psi'(z)}{\psi(z)}$$

Here $\psi'(z)/\psi(z)$ is analytic at $z = b_k$. Now applying the residue theorem to $f'(z)/f(z)$ over the region bounded by C, we have

$$\oint_C \frac{f'(z)}{f(z)}dz = 2\pi i \sum \text{residues}$$

$$= 2\pi i \left(\sum n_k - \sum p_k\right)$$

$$= 2\pi i(N-P)$$

where N is the number of zeros and P is the number of poles of $f(z)$ within C, each counted as many times as its multiplicity.

Thus

$$\frac{1}{2\pi i}\oint_C \frac{f'(z)}{f(z)}dz = N - P$$

An alternative form of this theorem is the following: We have

$$N - P = \frac{1}{2\pi i}\oint_C \frac{f'(z)}{f(z)}dz = \frac{1}{2\pi i}\oint_C d(\ln f(z))dz$$

Performing integration,

$$N - P = \frac{1}{2\pi i}[\ln f(z)]_C$$

or

$$N - P = \frac{1}{2\pi i}[\ln |f(z)| + i \arg f(z)]_C$$

$$= \frac{1}{2\pi}[\arg f(z)]_C$$

$$= \frac{1}{2\pi}[\text{variation of arg } f(z) \text{ around } C]$$

The statement

$$\frac{1}{2\pi}(\text{increase in arg } f(z) \text{ around } C) = N - P$$

is known as the **argument principle**. This information is of great value in finding zeros and poles of analytic functions.

Remark

If $f(z)$ is analytic within and on a closed contour C and does not vanish on C, then the number of zeros of $f(z)$ within C is equal to $1/2\pi$ times the net variation in the argument of $f(z)$ as z traverses the curve C in the counter-clockwise direction.

Thus when $f(z)$ does not have any poles, i.e. $P = 0$,

$$N = \text{number of zeros of } f(z)$$
$$= \frac{1}{2\pi}[\text{variation of arg } f(z) \text{ around } C]$$

In geometric terms, this **argument principle** means that if the locus of $w = f(z)$ is plotted for values ranging around the given circle C, then the number of times this locus encircles the origin in the w-plane is the number of zeros of $f(z)$ within C.

Figure 5.19(a) reveals that under the mapping of $w = f(z)$, all zeros of the function $f(z)$ will be mapped onto the origin of the w-plane. Thus to determine whether or not each of the roots of a polynomial equation $f(z) = 0$ has a negative real part, we proceed as follows:

In the z-plane let the contour C consist of the segment of the imaginary axis between $-R$ to R (R goes to infinity) together with the semi-circle lying in the right half plane and having this segment as diameter as shown in Fig. 5.19(b). Since a polynomial equation has only a finite number of roots, any roots of $f(z) = 0$ which lie in the right half plane, i.e. any roots which have positive real parts, will also lie within C. Thus by means of the transformation $w = f(z)$, let z range over the contour C (semi-circle and the imaginary axis) and let the locus of the corresponding values be plotted. If the image curve Γ does not encircle the origin in the w-plane, then we are sure that $f(z) = 0$ has no roots which have positive real parts in the right half plane.

The labour of plotting the image curve in the w-plane can be reduced considerably by letting R tend to infinity.

Let $z = Re^{i\theta}$ for $0 \le \theta \le 2\pi$ or $-\pi \le \theta \le \pi$ so $w = f(z) = f(Re^{i\theta})$, thus $f(z)$ can be expanded as a polynomial of degree n.

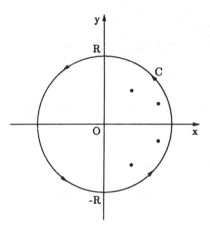

Fig. 5.19(a): Argument principle.

But

$$
\begin{aligned}
f(z) &= a_0 z^n + a_1 z^{n-1} + \cdots + a_n \\
&= a_0 (Re^{i\theta})^n + a_1 (Re^{i\theta})^{n-1} + \cdots + a_{n-1}(Re^{i\theta}) + a_n \\
&= R^n \left[a_0 e^{in\theta} + \frac{a_1}{R} e^{i(n-1)\theta} + \cdots + \frac{a_{n-1}}{R^{n-1}} e^{i\theta} + \frac{a_n}{R^n} \right]
\end{aligned}
$$

When $R \to \infty$, $w = f(z) \approx a_0 R^n e^{in\theta}$ for $-n\pi \le n\theta \le n\pi$ or $-n\pi \le \phi \le n\pi$ where $\phi = \arg f(z) = \arg w$. That means when θ varies from $-\pi$ to π, ϕ varies from $-n\pi$ to $n\pi$.

Now referring to Fig. 5.19(b), we can conclude that when θ varies from $-\pi/2$ to $\pi/2$ in the z-plane, the ϕ varies from $-n\pi/2$ to $n\pi/2$ in the w-plane.

5.8.1   The Nyquist criterion

Let us consider a differential equation of order n with constant coefficients:

$$
a_0 \frac{d^n y}{dt^n} + a_1 \frac{d^{n-1} y}{dt^{n-1}} + \cdots + a_n y = f(t) \tag{5.82}
$$

The initial conditions are

$$
t = 0 : y_0 = y_0' = \cdots = y_0^{(n)} = 0 \tag{5.83}
$$

Taking the Laplace transform of (5.82) and using the initial conditions (5.83) we obtain

$$
(a_0 s^n + a_1 s^{n-1} + \cdots + a_n)\mathcal{L}\{y(t)\} = \mathcal{L}\{f(t)\}
$$

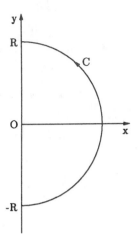

Fig. 5.19(b): Argument principle.

We write $\psi(s) = a_0 s^n + a_1 s^{n-1} + \cdots + a_n$. Therefore

$$\mathcal{L}\{y(t)\} = \frac{1}{\psi(s)}\mathcal{L}\{f(t)\} \qquad (5.84)$$

We know

$$\mathcal{L}\{y(t)\} = \frac{\mathcal{L}\{f(t)\}}{\psi(s)} = \frac{P(s)}{Q(s)}$$

where $\phi(s) = 1/\psi(s)$ is the transfer function and $\phi(i\omega) = 1/\psi(i\omega)$ is the frequency response function. Consider that the denominator of $\mathcal{L}\{y(t)\}$ is equal to $Q(s)$.

Let $Q(s) = a_0 s^n + a_1 s^{n-1} + \cdots + a_n$ be a polynomial of degree n. $Q(s)$ is an analytic function of the complex variable s, then the number N of zeros of $Q(s)$ in a region bounded by a closed contour C can be found by the argument principle

$$N = \frac{1}{2\pi i} \oint_C \frac{Q'(s)}{Q(s)} ds$$

$$= \frac{1}{2\pi}(\text{total increment in arg } Q(s) \text{ around } C) \qquad (5.85)$$

To study stability we wish to determine whether $Q(s)$ has any zeros in the right half plane, $Re(s) \geq 0$.

Therefore we choose the following semicircle of radius R as the closed contour C:

$$Re(s) = 0, \quad -R \leq Im(s) \leq R; \quad |s| = R, \quad -\frac{\pi}{2} \leq \arg s \leq \frac{\pi}{2}$$

We assume R to be sufficiently large such that all the roots of $G(s)$ will be inside the contour C.

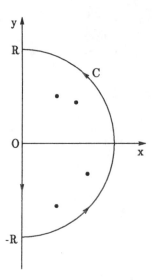

Fig. 5.20: Application of argument principle to stability.

Now let s range over the contour C, and in an auxiliary w-plane let the locus of the corresponding values of $w = Q(s)$ be plotted. If the curve does not enclose the origin in the w-plane, then according to the argument principle, $Q(s) = 0$ has no roots in the right half plane, and the solution of the corresponding differential equation will be stable. If, further, this curve does not pass through the origin in the w-plane, then $Q(s) = 0$ has no roots on the imaginary axis, i.e. all roots of $Q(s) = 0$ have negative real parts, and consequently the solution of the differential equation is stable. On the other hand, if the curve $w = Q(s)$ encircles the origin in the w-plane, a net number of times k, then $Q(s) = 0$ has k roots in right half plane, i.e. it has k roots with positive real parts and accordingly the solution of the differential equation is **unstable**. It is to be noted that if Q has a zero on the imaginary axis, then arg Q becomes undefined at this point, because of the roots with $Re(s) = 0$, then the system must be unstable. It is impossible to know how large R must be chosen to ensure that all roots of $Q(s)$ in the right half plane lie inside C. We answer this question by the following theorem.

Theorem 5.25

The polynomial $Q(s) = a_0 s^n + a_1 s^{n-1} + \cdots + a_n$ has no roots for which $|s| \geq R$, if $R = 1 + M/|a_0|$ and M is the largest of $|a_0|, \cdots, |a_n|$.

Proof

Let s be a root and let $|s| > 1$. Then

$$a_0 s^n + a_1 s^{n-1} + \cdots + a_n = 0$$

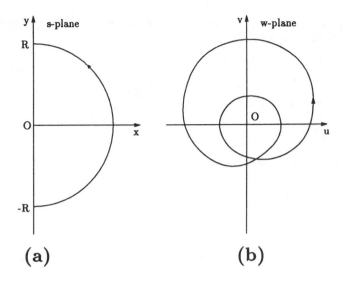

Fig. 5.21: Nyquist diagram.

$$a_0 s^n \left[1 + \frac{a_1}{a_0 s} + \frac{a_2}{a_0 s^2} + \cdots + \frac{a_n}{a_0 s^n} \right] = 0$$

Then

$$1 = -\frac{1}{a_0 s} \left(a_1 + \frac{a_2}{s} + \cdots + \frac{a_n}{s^{n-1}} \right)$$

so that

$$1 \leq \frac{1}{|a_0||s|} \left(|a_1| + \frac{|a_2|}{|s|} + \cdots + \frac{|a_n|}{|s^{n-1}|} \right)$$

$$\leq \frac{M}{|a_0||s|} \left(1 + \frac{1}{|s|} + \cdots + \frac{1}{|s^{n-1}|} \right)$$

$$1 \leq \frac{M}{|a_0||s|} \sum_{k=0}^{\infty} \frac{1}{|s|^k} = \frac{M}{a_0} \frac{1}{|s| - 1}$$

In the last step we took advantage of the fact that the geometric series $\sum |s|^{-k}$ converges for $|s| > 1$ and can be summed by the familiar rule.

Thus we have finally $1 < (M/a_0)\{1/(|s| - 1)\}$ or $|s| < 1 + M/|a_0|$ as asserted.

In computing the total change of arg Q on C we use the fact that as s traverses C, $w = Q(s)$ passes a curve C_w in the w-plane. The number of zeros inside C is then equal to the number of times C_w encircles the origin of the w-plane in the positive direction. Hence it is sufficient to graph C_w and determine whether it encircles $w = 0$.

In graphing C_w we can simplify the work by noting that along the circular part of C, $Q(s)$ is approximately equal to

$$a_0 s^n \approx a_0 R^n e^{in\theta}, \text{ for large } R$$

Hence this part of C corresponds to an approximately circular arc in the w-plane along which $\arg w$ varies from $-n\pi/2$ to $n\pi/2$.

Example 2

Find the zeros of $Q(s)$ which are in the right half plane where

$$Q(s) = s^2 + 3s - 4$$

Solution

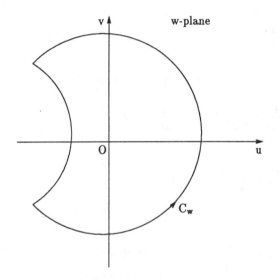

Fig. 5.22: Stability plot.

Here, of course, we immediately verify that the system is unstable. Because

$$s^2 + 3s - 4 = 0$$

and therefore $s = -4, 1$. Here $s = 1$ is in the right half plane of C and so the system is unstable. By the above diagram (Fig. 5.22), we then have along the diameter of C, $Q = -4 - \omega^2 + 3i\omega$, where $u = -4 - \omega^2$ and $v = 3\omega$ and the path $W = Q(s)$ is easily traced.

Along the circular part, $w = s^2$ approximately, and accordingly we trace the circle $|w| = R^2$, $-\pi < \arg w < \pi$. In accordance with the previous theorem, R is chosen as 5. The resulting curve C_w encircles the origin $w = 0$ once; and hence there is one zero inside C, and the system is unstable.

Example 3

Find the zeros of $Q(s)$ which are in the right half plane where

$$Q(s) = s^2 + 3s + 2 = (s+1)(s+2)$$

Solution

The roots of $Q(s) = 0$ are given by $s = -1, -2$. Both the roots are in the left half plane. Therefore, the system is **stable**.

Solution by the following diagram (Fig. 5.23):

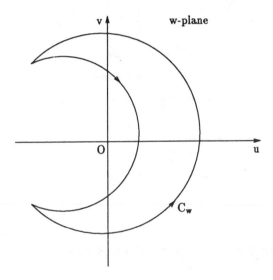

Fig. 5.23: Stability plot.

Along the diameter of C, $(s = i\omega)$ and therefore $Q = 2 - \omega^2 + 3i\omega$ and the path $w = Q(s)$ is easily traced. Here, $u = 2 - \omega^2$, $v = 3\omega$.

Along the circular part $w = s^2 \approx R^2$, $-\pi < \arg w < \pi$. Here R is chosen 4. The resulting curve C_w encircles the origin $w = 0$ not a single time. Hence there is no zero inside C, i.e. the system is **stable**. A diagram such as that of Figs. 5.22 and 5.23 is known as a Nyquist diagram, and determination of stability by determining how many times C_w encircles $w = 0$ is known as the **Nyquist criterion**.

Example 4

If

$$\mathcal{L}\{y(t)\} = \frac{s+1}{s^3 + s^2 + 9s + 1}$$

then determine by the Nyquist criterion whether the system is stable or not.

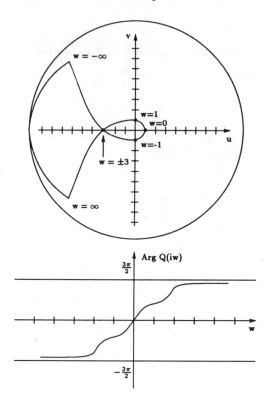

Fig. 5.24(a): Plot of $Q(s) = s^3 + s^2 + 9s + 1$ and $\arg Q(s)$ for $s = iw$.

Solution

Here

$$Q(s) = s^3 + s^2 + 9s + 1$$

By the test stated in Theorem 5.22, we can conclude that since $a_1 a_2 - a_0 a_3 = 8$ which is greater than zero, all the roots must be in the left half plane. Thus the system is stable.

Let us confirm this by the Nyquist criterion. We begin by plotting

$$w = Q(s) = s^3 + s^2 + 9s + 1$$

for values of s on the imaginary axis (diameter), i.e. for $s = iw$ and w ranging from $-\infty$ to ∞.

The parametric equations for the image curve can be obtained as:

$$Q(iw) = -iw^3 - w^2 + 9iw + 1$$

and so the real and imaginary parts are $u = 1 - \omega^2$ and $v = 9\omega - \omega^3$.

Figure 5.24(a) shows the plot of this curve together with the plot of $\arg Q$. Evidently as s traverses the imaginary axis from $i\infty$ to $-i\infty$, $\arg Q$ varies from $3\pi/2$ to $-3\pi/2$. The net variation is -3π contributed by the imaginary axis. The net variation contributed by the semi-circular portion of the contour C is the value $n\pi = 3\pi$. Adding these two variations gives a total net variation of zero as the entire contour C is traversed.

Hence $Q(s)$ has no zeros in the right half of the s-plane. Moreover, the image curve C_w encircles the origin $w = 0$ not a single time. Therefore, the inverse $y(t)$ is **stable**.

Example 5

Discuss the stability of $y(t)$ if

$$\mathcal{L}\{y(t)\} = \frac{s^2 + 1}{s^3 + s^2 + s + 9}$$

Solution

Here

$$Q(s) = s^3 + s^2 + s + 9$$

By the test stated in Theorem 5.22, we can conclude that since $a_1 a_2 - a_0 a_3 = -8$ which is less than zero, some roots must be in the right half plane. Thus the system is **unstable**.

Let us confirm this by the Nyquist criterion. We begin by plotting

$$w = Q(s) = s^3 + s^2 + s + 9$$

for values of s on the imaginary axis (diameter), i.e. for $s = i\omega$ and ω ranging from $-\infty$ to ∞. The parametric equations for the image curve can be obtained as:

$$Q(i\omega) = -i\omega^3 - \omega^2 + i\omega + 9$$

and so the real and imaginary parts are $u = 9 - \omega^2$ and $v = \omega - \omega^3$. Figure 5.24(b) shows the plot of this curve together with the plot of $\arg Q$. Evidently as s traverses the imaginary axis from $i\infty$ to $-i\infty$, $\arg Q$ varies from $3\pi/2$ to $5\pi/2$. The net variation is π contributed by the imaginary axis. The net variation contributed by the semi-circular portion of the contour C is the value $n\pi = 3\pi$. Adding these two variations gives a total net variation of 4π as the entire contour C is traversed. Dividing this by 2π, we obtain 2 as the number of zeros of $Q(s)$ in the right half plane. Moreover, the image curve C_w encircles the origin two times.

Therefore, the inverse $y(t)$ is **unstable**.

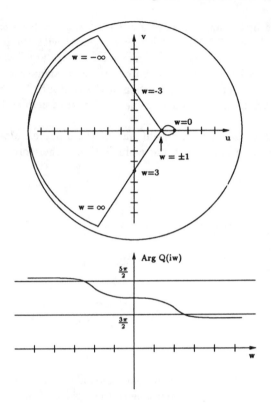

Fig. 5.24(b): Plot of $Q(s) = s^3 + s^2 + 9s + 1$ and $\arg Q(s)$ for $s = i\omega$.

5.8.2   Applications

The Nyquist stability criterion finds its best known application in the stability analysis of closed loop control systems. One common problem in engineering is to make the output $y(t)$ of a system follow quickly and accurately changes made in the input $f(t)$ to the system.

In an open loop system (Fig. 5.25a) that is often difficult to accomplish, specially prolonged oscillation of $y(t)$ about its desired value may well follow an abrupt change of the input $f(t)$ to some desired new value. One possible way to remedy this situation is to construct a **feedback loop** like the one shown in Fig. 5.25b, to sample the output and feed it back to a differential device which will in turn transmit the **error signal** $f(t) - y(t)$ as a modified or a corrected input to the original system. More generally, the output $y(t)$ may be modified by some additional device in the feedback loop to produce the feedback signal $y_f(t)$ before it is fed to the differential device (Fig. 5.25c).

With reference to Fig. 5.25b, we can write

$$\mathcal{L}\{y(t)\} = \phi(s)[\mathcal{L}\{f(t)\} - \mathcal{L}\{y(t)\}]$$

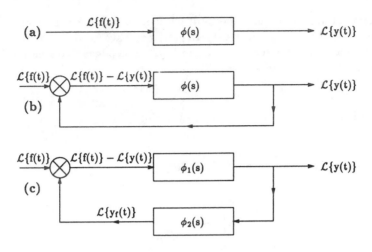

Fig. 5.25: Systems with feedback loop.

so that we obtain

$$\mathcal{L}\{y(t)\} = \frac{\phi(s)}{1 + \phi(s)}\mathcal{L}\{f(t)\}$$

Thus $\mathcal{L}\{y(t)\} = \phi_1(s)\mathcal{L}\{f(t)\}$ where $\phi_1 = \phi(s)/\{1 + \phi(s)\}$ is the final transfer function. Here, $\phi(s)$ is the open loop transfer function, and $\phi_1(s)$ is the closed loop transfer function.

In Fig. 5.25c, let $\phi_1(s)$ and $\phi_2(s)$ be the transfer functions of the original system and the feedback loop, respectively. Then, from the definition of a transfer function (ratio of the transformed output to the transformed input) we can write

$$\mathcal{L}\{y(t)\} = \phi_1(s)[\mathcal{L}\{f(t)\} - \mathcal{L}\{y_f(t)\}]$$

$$\mathcal{L}\{y_f(t)\} = \phi_2(s)\mathcal{L}\{y(t)\}$$

If we eliminate $\mathcal{L}\{y_f(t)\}$ between these two equations, we obtain

$$\mathcal{L}\{y(t)\} = \frac{\phi_1(s)}{1 + \phi_1(s)\phi_2(s)}\mathcal{L}\{f(t)\}$$

Evidently $\phi_1(s)/\{1 + \phi_1(s)\phi_2(s)\}$ is the overall transfer function of the entire closed loop system.

The equation of the stability of a feedback system can be answered by an examination of the Laplace transform of the output, namely,

$$\frac{\phi_1(s)}{1 + \phi_1(s)\phi_2(s)}\mathcal{L}\{f(t)\}$$

If the original system without the feedback loop is stable for the input $f(t)$, as we shall suppose, then the product $\phi_1(s)\mathcal{L}\{f(t)\}$ can have no poles in the right half of the s plane and the stability of the overall system depends solely on the location of the zero of the denominator

$$1 + \phi_1(s)\phi_2(s)$$

Hence as before, we plot the locus of the function

$$w = 1 + \phi_1(s)\phi_2(s)$$

as s ranges over the contour C of Fig. 5.26.

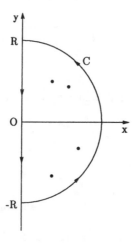

Fig. 5.26: Contour C in s-plane.

Since $\phi_1(s)$ and $\phi_2(s)$ are themselves Laplace transforms, each approaches zero when $s = R \to \infty$. Thus $w = 1$ as $R \to \infty$ and hence contributes nothing to the total variation of arg w as s traverses C.

Thus, to determine stability, it is necessary only to plot

$$w = 1 + \phi_1(s)\phi_2(s)$$

for values of s on the imaginary axis and determine whether or not the resulting curve encloses the origin. The curve, however, can be constructed simply by plotting $1+\phi_1(i\omega)\phi_2(i\omega)$ for positive values of ω and then reflecting the resulting arc in the real axis. In practice, instead of plotting $w = 1 + \phi_1(i\omega)\phi_2(i\omega)$ and observing whether the image curve encircles the origin, it is customary to plot $w = \phi_1(s)\phi_2(s)$ and observe whether it encircles the point $w = -1$. The equivalence of these two method is obvious. This type of study can generally be found in a large number of books such as *Analysis and Design of Feedback Control Systems* by G.J. Thaler and R.G. Brown, 2nd Ed., McGraw-Hill, New York, 1960, and *Servo-mechanisms and Regulating System Design* by H. Chesnut and R.W. Meyer, Wiley, New York, 1951.

Exercises

1. Find the image of the closed contour C (Fig. 5.26) under the transformation $w = s^3 + 1$ for the cases $R = 1, 2, 3$. Use these graphs to determine the number of roots of $s^3 + 1 = 0$ in the right half plane.

2. Verify that $w = s^2 + 1$ has one zero inside the circle $|s - i| = 1$ and two zeros inside the circle $|s| = 2$, by plotting w as s varies around the respective circles.

   Graph $\arg Q(iw)$ against w for $-\infty < w < \infty$ and use the results to determine stability for the following choices of Q:

3. $Q(s) = s^3 + s + 9$

4. $Q(s) = s^3 + 2s^2 + 3s + 1$

5. $Q(s) = s^4 + s^3 + 7s^2 + 4s + 10$

6. Show that all the roots of the equation $s^7 - s^3 + 12 = 0$ are located in the annulus bounded by the circle $|s| = 1$ and $|s| = 2$.

7. Show that if an electric system has an open-loop transfer function

$$\phi(s) = \frac{s^2 + s + 7}{s^3 + 2s^2 + s + 1}$$

then the system is stable; however the closed-loop system is unstable.

Chapter 6

Conformal mapping

6.1 Introduction

A geometrical representation of a real-valued function of a real variable such as $y = y(x)$ can be easily demonstrated by drawing graphs. But, the geometrical representation for $w = f(z)$ when z varies over the entire complex plane cannot be plotted in the Cartesian fashion because, in the complex domain, a relation $w = f(z)$, defined for all z, i.e. $u + iv = f(x + iy)$ involves four real variables, namely two independent variables x and y and the two dependent variables u and v which constitute four-dimensional space. Therefore, to avoid this difficulty, it will be simpler if we consider the two planes, namely z and w separately.

Consider then the two planes, one the z plane in which the point $z = x + iy$ is to be plotted, and the other the w plane in which the point $w = u + iv$ is to be plotted. Then a function $w = f(z)$ will be represented by a correspondence between the two points of these two Cartesian planes. As, for example, whenever a curve is drawn in the z plane, the function $w = f(z)$ determines the corresponding curve in the w plane. When a function $w = f(z)$ is thought of in this way, it is often referred to as a **mapping** or a **transformation**. Thus, the function $w = f(z)$ defines a **mapping** or a **transformation** of the z plane into the w plane.

Example 1

The mapping $w = z^2$ transforms a straight line in the z plane into a parabola in the w plane. Discuss it.

Solution

From the given transformation we have $u + iv = x^2 - y^2 + 2ixy$.

Equating real and imaginary parts, we obtain

$$\left. \begin{array}{l} u = x^2 - y^2 \\[2mm] v = 2xy \end{array} \right\} \tag{6.1}$$

Let

$$y = mx + c \tag{6.2}$$

be a line in the z plane. Then, from (6.1) we have

$$\left. \begin{array}{l} u = x^2 - (mx + c)^2 = (1 - m^2)x^2 - 2mxc - c^2 \\ \\ v = 2x(mx + c) = 2mx^2 + 2cx \end{array} \right\} \tag{6.3}$$

and

Now, if we regard these equations in (6.3) as two simultaneous equations in x and x^2, we can solve, getting

$$\left. \begin{array}{rcl} x^2 & = & (u + c^2 + mv)/(1 + m^2) \\ \\ x & = & \{v(1 - m^2) - 2m(u + c^2)\}/\{2c(1 + m^2)\} \end{array} \right\} \tag{6.4}$$

Hence, we obtain

$$\left(\frac{u + c^2 + mv}{1 + m^2} \right) = \left\{ \frac{v(1 - m^2) - 2m(u + c^2)}{2c(1 + m^2)} \right\}^2$$

and, after simplification, yields

$$\begin{array}{l} (1 - m^2)^2 v^2 + 4m^2 u^2 - 4m(1 - m^2)uv \\ \qquad - 4c^2(1 - m^2)u - 8mc^2 v = 4c^4 \end{array} \tag{6.5}$$

which is the equation of a parabola. For simplicity, when $m = 1$ and $c = 0$, eqn (6.5) reduces to $u^2 = 0$, and for $m = 0$ and $c = 0$, it reduces to $v^2 = 0$. Graphically then, we obtain

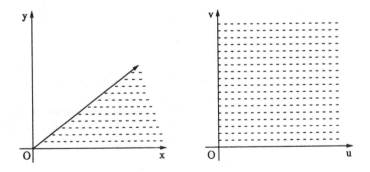

Fig. 6.1: Conformal mapping.

6.2   Some simple mapping

Consider the transformation $w = f(z) = z^2$ such that $u + iv = x^2 - y^2 + 2ixy$ where $u = x^2 - y^2$ and $v = 2xy$. These are the equations of the transformation from z plane to w plane. Now, consider for simplicity $x = \alpha$ which are the lines parallel to y axis for different values of the constant α. Then we have $u = \alpha^2 - y^2$ and $v = 2\alpha y$ and eliminating the parameter y, we obtain $u = \alpha^2 - v^2/(4\alpha^2)$, which is a parabola and for different α we can form a system of parabolas. On the other hand, consider $y = \beta$ which are the lines parallel to the x axis for different values of β. Then, we have $u = x^2 - \beta^2$ and $v = 2\beta x$ and eliminating the parameter x, we obtain $u = (v^2/4\beta^2) - \beta^2$, which is again a parabola and for different β we can form a system of parabolas.

Thus, under the transformation $w = z^2$, a straight line $x = \alpha$ in the z plane transforms to $u = \alpha^2 - (v^2/4\alpha^2)$ in the w plane. Similarly, a straight line $y = \beta$ in the z plane transforms to $u = (v^2/4\beta^2) - \beta^2$. In graphical representation then, for fixed values of $\alpha = 1, 2$ and $\beta = 1, 2$, the following Fig. 6.2 shows the mapping:

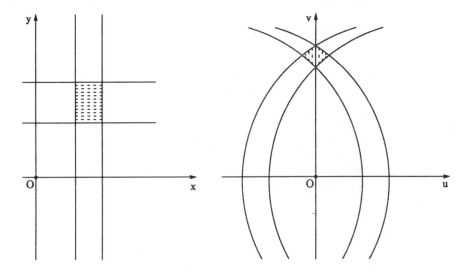

Fig. 6.2: Mapping by $w = z^2$ from z plane to w plane.

Also, the mapping from w plane to z plane can be determined very easily. For instance, consider $u = k$ such that $x^2 - y^2 = k$ and $v = c$ such that $xy = c/2$.

In graphical representation, Fig. 6.3 is then obtained.

Exercises

1. Determine the angle of rotation at the point $z = 1 + i$ for the transformation in Example 1.

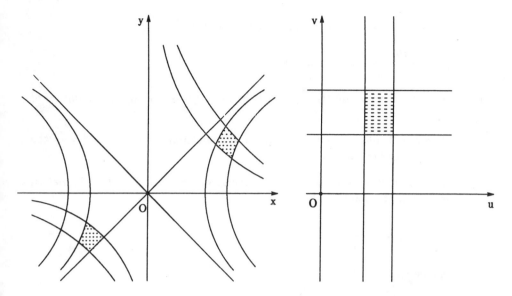

Fig. 6.3: Mapping by $w = z^2$ from w plane to z plane.

2. Determine the image of the circle $|z| = a^2$ under the mapping in Exercise 1.

3. Discuss the transformation defined by $w = 1/z$.

4. Determine the angle of rotation produced by the transformation in Exercise 3:

   (a) at the point $z = 2$
   (b) at the point $z = 2i$

5. Discuss the transformation defined by $w = (x^2 - y^2) + 2ixy$. Determine its significant difference from the transformation

   $$w = (x^2 - y^2) + 3ixy$$

6. Discuss the transformation defined by $w = z^3$. Plot the image of the line $v = 1$. Find the equation of the image of the line $y = 1$.

7. Show that the transformation $w = e^z$ is everywhere conformal.

8. Find the equations of the transformation defined by the function

   $$w = \frac{z - i}{z}$$

   and show that every circle through the origin in the z plane is transformed into a straight line.

9. Show that $w = \sin z$ is conformal at all points except

$$z = \frac{(2n-1)\pi}{2}, \quad (n = 0, \pm1, \pm2, \cdots)$$

10. In Exercise 8, what are the images of the lines $x = \alpha$? What are the images of the lines $y = \beta$?

11. In Exercise 9, what is the image in the fundamental strip $-\pi < y \leq \pi$ of the line $u = c_1$ and of the line $v = c_2$?

12. Show that if $z = g(w)$ is the local inverse of the conformal transformation $w = f(z)$ at the point z_0, then

$$g'(w_0) = \frac{1}{f'(z_0)}$$

where $w_0 = f(z_0)$.

13. Determine the local inverse of the transformation $w = e^z$ at the point

   (a) $z_0 = 0$

   (b) $z_0 = \pi i$

Note

If $w = f(z)$ is analytic in a given region such that the derivative of $f(z)$ exists in that region and that $f'(z)$ is different from zero, then this transformation is called **conformal**. In the following we discuss more rigorously the concept of **conformal mapping**.

6.3 Conformal mapping

We have seen in the last section that the set of equations

$$u = u(x, y)$$

$$v = v(x, y)$$

$$(6.6)$$

defines, in general, a transformation or mapping which establishes a correspondence between points in the uv and xy planes. The eqns (6.6) are called transformation equations. This mapping or transformation is said to be one to one if to each point of the uv plane there corresponds one and only one point of the xy plane, and conversely. In that situation, a set of points in the xy plane is mapped into a set of points in the uv plane and conversely. These corresponding sets of points in the two planes are often called images of each other.

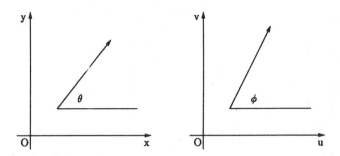

Fig. 6.4: Δz and Δw under a mapping of $w = f(z)$.

Solving for x and y in (6.6), we obtain

$$\begin{aligned} x &= x(u, v) \\ y &= y(u, v) \end{aligned} \qquad (6.7)$$

This transformation is often called inverse transformation. The Jacobian of transformation of (6.6) is then defined as

$$J\left(\frac{u, v}{x, y}\right) = \frac{\partial(u, v)}{\partial(x, y)} = \begin{vmatrix} \partial u/\partial x & \partial u/\partial y \\ \partial v/\partial x & \partial v/\partial y \end{vmatrix} \qquad (6.8)$$

The Cauchy-Riemann conditions for the analytic function $w = f(z)$ or $u + iv = f(x + iy)$ are given by

$$\left. \begin{aligned} \partial u/\partial x &= \partial v/\partial y \\[2mm] \partial u/\partial y &= -\partial v/\partial x \end{aligned} \right\} \qquad (6.9)$$

Now, using (6.9) into (6.8), we obtain

$$\begin{aligned} J\left(\frac{u, v}{x, y}\right) &= \begin{vmatrix} \partial u/\partial x & -\partial v/\partial x \\ \partial v/\partial x & \partial u/\partial x \end{vmatrix} \\[3mm] &= \left(\frac{\partial u}{\partial x}\right)^2 + \left(\frac{\partial v}{\partial x}\right)^2 \\[3mm] &= \left| \frac{\partial u}{\partial x} + i\frac{\partial v}{\partial x} \right|^2 = |f'(z)|^2 \end{aligned}$$

The condition that there exists a one to one correspondence between the transformations (6.6) and (6.7), is that the Jacobian determinant is different from zero. It is worth mentioning here that the non-vanishing of a Jacobian ensures that the mapping is locally one-to-one. That means $f'(z) \neq 0$. Points where $f'(z) = 0$ are called critical points.

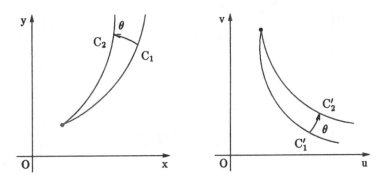

Fig. 6.5: A sketch of conformal mapping.

From the above discussion, we establish that if $f(z)$ is analytic, the function $w = f(z)$ will have a single-valued inverse in the neighbourhood of the critical point. Now, consider a value of z and its image $w = f(z)$, where $f(z)$ is analytic. Let $\Delta z = |\Delta z| e^{i\theta}$ and $\Delta w = |\Delta w| e^{i\phi}$, the corresponding increments of these quantities (see Fig. 6.4).

Then

$$f'(z) = \lim_{\Delta z \to 0} \frac{\Delta w}{\Delta z} = \lim_{\Delta z \to 0} \frac{|\Delta w| e^{i\phi}}{|\Delta z| e^{i\theta}} = \lim_{\Delta z \to 0} \left| \frac{\Delta w}{\Delta z} \right| e^{i(\phi - \theta)}$$

From this it is clear that $|f'(z)| = \lim_{\Delta z \to 0} |\Delta w / \Delta z|$ and $\arg f'(z) = \lim_{\Delta z \to 0} (\phi - \theta)$.

This implies that to any degree of approximation

$$|\Delta w| = |f'(z)| . |\Delta z| \tag{6.10}$$

and

$$\phi = \theta + \arg f'(z)$$

or

$$\arg(\Delta w) = \arg(\Delta z) + \arg f'(z) \tag{6.11}$$

The fact that $f'(z)$ exists means that both $|f'(z)|$ and $\arg f'(z)$ are independent of the path through which Δz approaches zero. They depend solely on z and not on the limiting orientation of the increment Δz. Thus, we conclude that in the mapping $w = f(z)$, the length of the infinitesimal segments, regardless of their direction, are altered by a factor $|f'(z)|$ which depends only on the point from which the segments are drawn.

It follows from (6.10) that since the infinitesimal lengths are magnified by the factor $|f'(z)|$, the infinitesimal areas are magnified by the factor $|f'(z)|^2$ which is exactly the Jacobian of the transformation $J(\frac{u, v}{x, y})$.

Similarly, we can conclude from (6.11) that the difference between the angles of an infinitesimal segment and its image is independent of the direction of the segment and depends only on the point from which the segment is drawn. In particular, two infinitesimal segments forming an angle will both be rotated in the same direction by the same amount. Thus, the measure of angle between them will in general be left invariant by the transformation. To see this point clearly, we refer to Fig. 6.5.

Let C_1 and C_2 be two smooth curves passing through the point z and let θ_1 and θ_2 be angles of inclination of a directed line tangent to C_1 and C_2, respectively at z. Then, according to (6.11) $\phi_1 = \theta_1 + \arg f'(z)$ and $\phi_2 = \theta_2 + \arg f'(z)$ are angles of inclination of the directed lines tangent to the image curves C_1' and C_2', respectively at the point $w_0 = f(z_0)$. Thus, $\phi_2 - \phi_1 = \theta_2 - \theta_1$; that is angle $\phi_2 - \phi_1$ from C_1' to C_2' is the same in magnitude and sense as the angle $\theta_2 - \theta_1$ from C_1 to C_2. This angle is denoted by θ. The following theorem results from the preceding discussion.

Theorem 6.1

If $f(z)$ is analytic and $f'(z) \neq 0$ is a region R, the mapping or transformation $w = f(z)$ is called **conformal** mapping at all points in R. However, a mapping that preserves the magnitude of each angle but not necessarily the sense is called an **isogonal** mapping.

If, however, $f'(z) = 0$, $\arg f'(z)$ is undefined and in this situation we cannot assert that the angles are preserved. To show this clearly, let us suppose that $f'(z)$ has an m-fold zero at $z = z_0$. Then, $f'(z)$ must contain the factor $(z - z_0)^m$, and so can be written as

$$f'(z) = (m+1)a(z - z_0)^m + (m+2)b(z - z_0)^{m+1} \cdots$$

where $a, b, \cdots$ are complex coefficients of no concern to us and the factor $(m+1), (m+2), \cdots$ have been inserted for convenience in integrating $f'(z)$ to obtain $f(z)$:

$$\int_{z_0}^{z} f'(z)dz = (m+1)a \int_{z_0}^{z} (z - z_0)^m dz + (m+2)b \int_{z_0}^{z} (z - z_0)^{m+1} dz + \cdots$$

or

$$f(z) - f(z_0) = a(z - z_0)^{m+1} + b(z - z_0)^{m+2} + \cdots$$

Let $f(z) - f(z_0) = \Delta w$ and $z - z_0 = \Delta z$. Then $\Delta w = a(\Delta z)^{m+1} + b(\Delta z)^{m+2} + \cdots$

or

$$\frac{\Delta w}{a(\Delta z)^{m+1}} = 1 + \frac{b}{a}(\Delta z) + \cdots$$

Taking the argument of both sides and letting $\Delta z \to 0$, we see that

$$\arg(\Delta w) - \arg(a(\Delta z)^{m+1}) = \arg(1) = 0$$

Thus we have $\arg(\Delta w) = \arg a + (m+1)\arg(\Delta z)$.

Now, let Δz_1 and Δz_2 be two infinitesimal segments which make an angle θ with each other, and let Δw_1 and Δw_2 be their images. Then, from the last equation, we have

$$\arg(\Delta w_1) = \arg a + (m+1)\arg(\Delta z_1)$$
$$\arg(\Delta w_2) = \arg a + (m+1)\arg(\Delta z_2)$$

Hence, subtracting one from the other, we have

$$\arg(\Delta w_2) - \arg(\Delta w_1) = (m+1)(\arg(\Delta z_2) - \arg(\Delta z_1))$$
$$= (m+1)\theta$$

Thus, we see that when $f'(z)$ has an m-fold zero at z_0, the measure of the angle is altered by the factor $(m+1)$. It can be easily seen that in the mapping defined by the analytic function $w = f(z)$, angles are generally preserved in magnitude and sense. In particular, when $m = 0$, i.e., $f''(z) \neq 0$,

$$\arg(\Delta w_2) - \arg(\Delta w_1) = \arg(\Delta z_2) - \arg(\Delta z_1) = \theta$$

As for instance, the mapping function $w = f(z) = z^2$ is everywhere analytic, and $f'(z) = 2z$ has a simple zero at $z = 0$. Referring to Fig. 6.5, the ray $\theta = c$ from the point $z = 0$ is mapped on the ray $\phi = 2c$ from the point $w = 0$. Thus, the angle between any two rays drawn from the critical point $z = 0$ is doubled by the transformation.

One important reason for studying conformal mapping is that solutions of Laplace's equation remain solutions of Laplace's equation when subjected to a conformal transformation. The following theorem asserts this statement clearly.

Theorem 6.2

If $\phi(x,y)$ is a solution of the equation

$$\frac{\partial^2 \phi}{\partial x^2} + \frac{\partial^2 \phi}{\partial y^2} = 0$$

then, when $\phi(x,y)$ is transformed into a function of u and v by conformal transformation, it will satisfy the equation

$$\frac{\partial^2 \phi}{\partial u^2} + \frac{\partial^2 \phi}{\partial v^2} = 0$$

everywhere except at the critical points.

Proof

Let $w = u(x, y) + iv(x, y) = f(z)$ define a conformal transformation by means of which $\phi(x, y)$ is transformed into a function of u and v. Then

$$\phi_x = \phi_u u_x + \phi_v v_x$$
$$\phi_y = \phi_u u_y + \phi_v v_y$$
$$\phi_{xx} = (\phi_{uu} u_x + \phi_{uv} v_x)u_x + \phi_u u_{xx}$$
$$+ (\phi_{vu} u_x + \phi_{vv} v_x)v_x + \phi_v v_{xx}$$
$$\phi_{yy} = (\phi_{uu} u_y + \phi_{uv} v_y)u_y + \phi_u u_{yy}$$
$$+ (\phi_{vu} u_y + \phi_{vv} v_y)v_y + \phi_v v_{yy}$$

Thus

$$\phi_{xx} + \phi_{yy} = \phi_{uu}(u_x^2 + u_y^2) + \phi_{vv}(v_x^2 + v_y^2)$$
$$+ 2\phi_{uv}(u_x v_x + u_y v_y) + \phi_u(u_{xx} + u_{yy})$$
$$+ \phi_v(v_{xx} + v_{yy})$$

Since $f(z)$ is analytic, and hence $u_{xx} + u_{yy} = 0$, and $v_{xx} + v_{yy} = 0$; and at the same time the Cauchy-Riemann conditions, $u_x = v_y$ and $u_y = -v_x$ must be satisfied. Thus, we have

$$\phi_{xx} + \phi_{yy} = (u_x^2 + v_x^2)(\phi_{uu} + \phi_{vv})$$
$$= |f'(z)|^2(\phi_{uu} + \phi_{vv})$$

Thus, at any point where the transformation is conformal, that is, where $f'(z) \neq 0$, $\phi_{xx} + \phi_{yy} = 0$ implies $\phi_{uu} + \phi_{vv} = 0$ as asserted.

Remark

This theorem confirms that if we need to solve a problem satisfying Laplace's equation with boundary conditions, within a complex region R, then it may be possible that R can be transformed to a simple region like a circle, or a half plane by conformal mapping in which Laplace's equation can be solved subject, of course, to the transformed boundary conditions. If this is the case, the resulting solution, when transformed back to R by inverse transformation, will be the required solution of the problem.

Example 1

Show that the transformation $w = f(z) = z^2$ is conformal at the point $z = 1 + i$, where the half lines $y = x$ $(y \geq 0)$ and $x = 1$ $(y \geq 0)$ intersect (see Fig. 6.6).

Solution

We denote those half lines by C_1 and C_2, with positive sense upwards, and observe that the angle from C_1 to C_2 is $\pi/4$ at their point of intersection (see Fig. 6.6). Since the image of a point (x,y) in the z plane is a point in the w plane whose rectangular coordinates are $u = x^2 - y^2$ and $v = 2xy$, the half line C_1 is transformed into the curve C_1' with parametric representation $u = 0$ and $v = 2y^2$ for the range $(0 \le y < \infty)$. This simply means that the image curve C_1' is the upper half of the v axis.

The half line C_2 is transformed into the curve C_2' represented by the equations $u = 1 - y^2$, $v = 2y$, for the range $(0 \le y \le \infty)$. Hence C_2' is the upper half of the parabola $v^2 = -4(u - 1)$. It is worth noting that in each case, the positive sense of the image curve is upward.

The slope of the image curve C_2' is given by

$$\frac{dv}{du} = \frac{dv/dy}{du/dy} = \frac{2}{-2y} = -\frac{2}{v}$$

Thus at the point $x = 1$ and $y = 1$, i.e at $u = 0$ and $v = 2$, $dv/du = -1$. Consequently, the angle from the image curve C_2' to the image curve C_1' at the point $w = f(1+i) = 2i$ is simply $\pi/4$. This is what is required by the conformity of the mapping at that point.

As evidenced, the angle of rotation $\pi/4$ at the point $z = 1 + i$ is a value of

$$\arg[f'(1 + i)] = \arg[2(1 + i)] = \frac{\pi}{4} + 2n\pi$$

where $n = 0, \pm 1, \pm 2, \dots$.

The scale factor at that point is the number $|f'(1 + i)| = |2(1 + i)| = 2\sqrt{2}$.

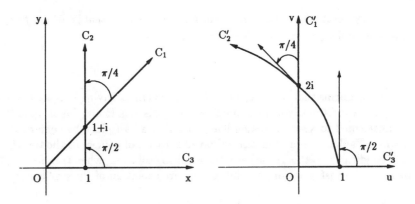

Fig. 6.6: Conformal mapping of $w = z^2$.

Exercises

1. Show that $u(x, y) = x^3 - 3xy^2$ is a harmonic function. Find a harmonic conjugate of this function and write the analytic function in terms of the complex variable z.

2. If $u = 2x^2 + y^2$ and $v = y^2/x$, show that the curves $u = $ constant and $v = $ constant are both orthogonal to each other, but the transformation is not conformal.

3. Consider the following transformations

$$u = u(x, y)$$
$$v = v(x, y)$$

Solving for x and y, we can write

$$x = x(u, v)$$
$$y = y(u, v)$$

Then show that

$$\frac{\partial x}{\partial u} = \left(\frac{1}{J}\right)\frac{\partial v}{\partial y}$$
$$\frac{\partial x}{\partial v} = -\frac{1}{J}\frac{\partial u}{\partial y}$$
$$\frac{\partial y}{\partial u} = -\frac{1}{J}\frac{\partial v}{\partial x}$$
$$\frac{\partial y}{\partial v} = \frac{1}{J}\frac{\partial u}{\partial x}$$

where

$$J = \frac{\partial(u, v)}{\partial(x, y)} = \text{a Jacobian determinant}$$

4. Verify the relations given in Exercise 3 for the following functions:

(a) $w = z + 2$
(b) $w = z^2$
(c) $w = z^3$

5. Show that

$$\frac{\partial(u, v)}{\partial(x, y)} = \frac{1}{\partial(x, y)/\partial(u, v)}$$

6. Verify the statement in Exercise 5 by the following functions:

(a) $w = z^2 - 1$

(b) $w = 2z + 1$

7. Find the area of the region into which the rectangle with vertices $z = 0, 2, 2 + i, i$ is transformed by the following functions:

     (a) $w = z^2$

     (b) $w = z^3$

8. Find the image of the hyperbola $x^2 - y^2 = 1$ under the transformation $w = \dfrac{1}{z}$.

6.4   The bilinear transformation

In the last section we discussed the conformal mapping or transformation. In this section, we shall discuss one of the most important conformal transformations known as the bilinear or linear fractional or Möbius transformations defined by the family of functions

$$w = \frac{az + b}{cz + d}, \quad ad \neq bc \tag{6.12}$$

The restriction $ad \neq bc$ is needed because if $ad = bc$, the fraction on the right side becomes independent of z and $w = k$ constant and the entire z-plane is mapped into the same point in the w-plane.

We shall consider 3 special cases before considering the general case. These special cases are:

                Case (I): $w = z + \alpha$
                Case (II): $w = \beta z$
                Case (III) $w = 1/z$

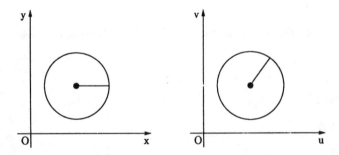

Fig. 6.7: Mapping of $w = z + \alpha$.

In Case (I), the mapping of the z-plane on to the w-plane by means of $w = z + \alpha$, where α is a complex constant, is a translation by means of the vector representing α. Thus, if $z = x + iy$ and $\alpha = \alpha_1 + i\alpha_2$, then the image of any point (x, y) in the z-plane is the point $(x + \alpha_1, y + \alpha_2)$ in the w-plane. In particular, we note that this rigid motion necessarily transforms circles into circles (Fig. 6.7).

Let circles in z-plane be given by $(x - a)^2 + (y - b)^2 = c^2$. Then, by the transformation $w = z + \alpha$, we obtain $u = x + \alpha_1$ and $v = y + \alpha_2$ which can be written as $u - \alpha_1 = x$ and $v - \alpha_2 = y$.

Then the circle in z-plane can be transformed as

$$(u - \alpha_1 - a)^2 + (v - \alpha_2 - b)^2 = c^2 \tag{6.13}$$

In Case (II), the mapping defined by $w = \beta z$, where β is complex, is readily obtained by using the polar forms of β and z. If $\beta = be^{i\phi}$ and $z = re^{i\theta}$, then $w = Re^{i\Psi} = br\, e^{i(\theta+\phi)}$. Thus,

$$R = br$$
$$\Psi = \theta + \phi$$

Therefore, any circles in the z-plane can be mapped into circles in the w-plane (see Fig. 6.8).

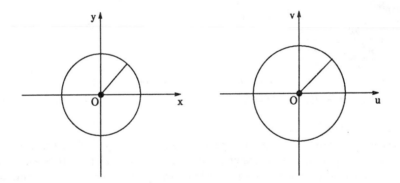

Fig. 6.8: Mapping by $w = \beta z$.

In Case (III), the mapping defined by $w = 1/z$ can be rewritten as

$$w = \frac{\bar{z}}{z\bar{z}} = \frac{\bar{z}}{|z|^2} \tag{6.14}$$

which shows that w is of length $1/|z|$ and has the direction of $\bar{z}$.

It can be easily seen that the image of a circle $|z| = \epsilon$ is the circle $|w| = 1/\epsilon$. This implies that a point very near to the origin in z-plane goes to the point very near to infinity.

Therefore the transformation $w = 1/z$ yields that

$$u + iv = \frac{1}{x + iy} = \frac{x - iy}{x^2 + y^2}$$

and comparing the real and imaginary parts:

$$u = \frac{x}{x^2 + y^2} \quad \text{and} \quad v = \frac{-y}{x^2 + y^2}$$

such that

$$u^2 + v^2 = \frac{1}{x^2 + y^2}$$

If a, b, c, and d are real numbers, the equation

$$a(x^2 + y^2) + bx + cy + d = 0 \tag{6.15}$$

represents any circle or line provided $a \neq 0$ or $a = 0$, respectively. Under the transformation $w = 1/z$, i.e.

$$u = \frac{x}{x^2 + y^2} \quad \text{and} \quad v = -\frac{y}{x^2 + y^2}$$

we obtain

$$a + \frac{bx}{x^2 + y^2} + \frac{cy}{x^2 + y^2} + \frac{d}{x^2 + y^2} = 0$$

or

$$a + bu - cv + d(u^2 + v^2) = 0 \tag{6.16}$$

which is a circle or line depending on $d \neq 0$ or $d = 0$.

However, any circle passing through the origin $(x = 0, y = 0)$ in the z-plane (i.e. $d = 0$) can be transformed to a straight line in the w-plane. Thus, with the aid of these three special transformations, we can construct the general bilinear transformation by using the following chain of special transformations:

$$w_1 \; = \; z + \frac{d}{c} \qquad c \neq 0$$

$$w_2 \; = \; cw_1 = cz + d$$

$$w_3 \; = \; \frac{1}{w_2} = \frac{1}{cz + d}$$

$$w_4 \; = \; \frac{bc - ad}{c} w_3 = \frac{bc - ad}{c(cz + d)}$$

and finally,

$$w = w_4 + \frac{a}{c} = \frac{bc - ad}{c(cz + d)} + \frac{a}{c} = \frac{az + b}{cz + d} \tag{6.17}$$

When $c = 0$, it is clear from the condition $ad - bc \neq 0$ that neither a nor d can be zero. Thus we have

$$w_1 = z + \frac{b}{a}$$

$$w = \frac{a}{d}w_1 = \frac{a}{d}\left(z + \frac{b}{a}\right) = \frac{az + b}{d} \tag{6.18}$$

which is simply a linear transformation.

Note

It is to be noted that each of these special transformations can be used to transform circles into circles or in special cases, straight lines to straight lines. The transformation (6.17) obtained from all these three is the general bilinear transformation.

The bilinear transformation defined in (6.17) depends on three essential constants, namely the ratios of any three of the coefficients a, b, c, d to the fourth. Hence, only three conditions are necessary to determine a bilinear transformation. In particular, the coordinates z_1, z_2 and z_3 of z have the image w_1, w_2 and w_3 of w which can determine a unique transformation. Indeed, the equation

$$\frac{(w - w_1)(w_2 - w_3)}{(w - w_3)(w_2 - w_1)} = \frac{(z - z_1)(z_2 - z_3)}{(z - z_3)(z_2 - z_1)} \tag{6.19}$$

defines such a transformation.

To establish a relation, we see that

$$w_i - w_j = \frac{az_i + b}{cz_i + d} - \frac{az_j + b}{cz_j + d} = \frac{(ad - bc)(z_i - z_j)}{(cz_i + d)(cz_j + d)}$$

Hence

$$\begin{aligned}
\frac{(w - w_1)(w_2 - w_3)}{(w - w_3)(w_2 - w_1)} &= \left\{ \frac{(ad - bc)(z - z_1)}{(cz + d)(cz_1 + d)} \times \frac{(ad - bc)(z_2 - z_3)}{(cz_2 + d)(cz_3 + d)} \right\} \div \\
&\quad\ \left\{ \frac{(ad - bc)(z - z_3)}{(cz + d)(cz_3 + d)} \times \frac{(ad - bc)(z_2 - z_1)}{(cz_2 + d)(cz_1 + d)} \right\} \\
&= \frac{(z - z_1)(z_2 - z_3)}{(z - z_3)(z_2 - z_1)} \tag{6.20}
\end{aligned}$$

The last fraction is defined as the **cross ratio**, or anharmonic ratio of the four numbers z, z_1, z_2, z_4. From this analysis, it is evident that the cross ratio of the four points is invariant under a bilinear transformation.

Theorem 6.3: Fundamental solution in logarithmic form

Any function which satisfies Laplace's equation

$$\frac{\partial^2 T}{\partial x^2} + \frac{\partial^2 T}{\partial y^2} = 0$$

$$\frac{\partial^2 S}{\partial x^2} + \frac{\partial^2 S}{\partial y^2} = 0 \qquad (6.21)$$

is called the harmonic function. Here $T(x,y)$ and $S(x,y)$ are harmonic functions. We have long since discovered that the real and the imaginary part of any analytic function $W = T + iS$ each satisfies Laplace's equation. Thus if we can find a function of z which is analytic in the upper half plane and whose real and imaginary parts take on the boundary values when z is real (i.e. along the x–axis), we are sure we shall have the required solution. We know that the **fundamental solution** of Laplace's equation in two-dimension is always in the form of $\ln r$ in real variables where r is the distance between the points (x,y) and (x_0, y_0). This concept can be extended easily to the complex plane by introducing an analytic function in the form $\ln(z - z_0)$ (which is analytic except at $z = z_0$) such that the real and imaginary parts of which satisfy Laplace's equation as mentioned above.

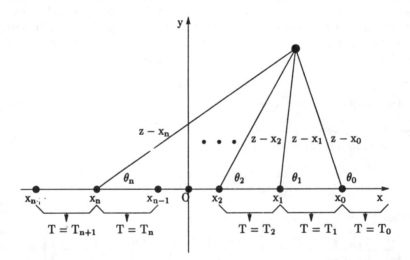

Fig. 6.9: The behaviour of $\arg(z - x_i)$ as z travels along the real axis.

Thus to obtain such a function which satisfies the boundary conditions along the real axis (this is called the Dirichlet Problem), we observe first that, (see Fig. 6.9), if $x_0 > x_1 > x_2 > \cdots > x_n$,

$$W = T_0 - i\frac{1}{\pi}[(T_1 - T_0)\ln(z - x_0) + (T_2 - T_1)\ln(z - x_1)$$
$$+ \cdots + (T_{n+1} - T_n)\ln(z - x_n)] \qquad (6.22)$$

is analytic everywhere in the upper half plane. Here W is usually defined as the complex potential. It is worth noting that since Laplace's equation is a linear partial differential equation, the general solution of this equation can be written as the sum of all possible logarithmic functions.

Hence eqn (6.22) is the most general solution which is analytic in the upper half plane. Its real part

$$T = T_0 + \frac{1}{\pi}[(T_1 - T_0)\arg(z - x_0) + (T_2 - T_1)\arg(z - x_1)$$
$$+ \cdots + (T_{n+1} - T_n)\arg(z - x_n)] \tag{6.23}$$

will be a solution of Laplace's equation everywhere in the upper half plane. Moreover, along the real axis, this solution takes on the boundary values shown in Fig. 6.9. This can be easily verified as described below:

If z is on the right-hand side of x_0, i.e. $z > x_0$, then

$$\arg(z - x_0) = 0$$
$$\arg(z - x_1) = 0$$
$$\cdots\cdots\cdots$$
$$\arg(z - x_n) = 0$$

and hence $T = T_0$ for the range $x_0 < x < \infty$. When z lies between x_1 and x_0, we have

$$\arg(z - x_0) = \pi$$
$$\arg(z - x_1) = 0$$
$$\arg(z - x_2) = 0$$
$$\cdots\cdots\cdots$$
$$\arg(z - x_n) = 0$$

Thus we obtain $T = T_0 + (1/\pi)[(T_1 - T_0)\pi] = T_1$ for $x_1 < x < x_0$.

Also when z lies between x_2 and x_1, then

$$\arg(z - x_0) = \pi$$
$$\arg(z - x_1) = \pi$$
$$\arg(z - x_2) = 0$$
$$\cdots\cdots\cdots$$
$$\arg(z - x_n) = 0$$

Thus we obtain $T = T_0 + (1/\pi)[(T_1 - T_0)\pi + (T_2 - T_1)\pi] = T_2$ for $x_2 < x < x_1$.

Continuing in this way we can show that the boundary value between $x_n < x < x_{n-1}$ can be derived as $T = T_n$.

Hence a function which is harmonic in the upper half plane $y > 0$, i.e. $Im(z) > 0$ and which takes on the prescribed values on the real axis x is given by eqn (6.23).

This equation can be written as

$$T(x,y) = T_0 + \frac{1}{\pi}\left[(T_1 - T_0)\arctan\left(\frac{y}{x - x_0}\right) + (T_2 - T_1)\arctan\left(\frac{y}{x - x_1}\right) + \cdots \right.$$
$$\left. + (T_{n+1} - T_n)\arctan\left(\frac{y}{x - x_n}\right)\right] \tag{6.24}$$

Theorem 6.4: Poisson's integral formula for the upper half plane

A function which is harmonic in the half plane $y > 0$ [i.e. $Im(z) > 0$] and which takes on the prescribed value $f(x)$ on the real axis [i.e. $T(x,0) = f(x)$, $-\infty < x < \infty$], is given by

$$T(x,y) = \frac{1}{\pi}\int_{-\infty}^{\infty}\frac{y}{(x - \eta)^2 + y^2}f(\eta)d\eta \tag{6.25}$$

This integral equation is called Poisson's integral equation. This integral can successfully handle the Dirichlet problems in two-dimensions for the half plane.

Proof

To prove this integral formula, we consider the following semi-circular closed contour. We know by Cauchy's integral formula, if $f(z)$ is an analytic function within and on the closed contour as shown in Fig. 6.10, then

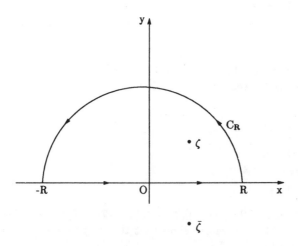

Fig. 6.10: A semi-circular closed contour.

$$f(\zeta) = \frac{1}{2\pi i}\oint_C \frac{f(z)}{z - \zeta}dz \tag{6.26}$$

where ζ is a point inside the contour.

Thus $\bar{\zeta}$ is the complex conjugate of ζ which is outside the given contour. Hence by Cauchy's integral formula

$$0 = \frac{1}{2\pi i} \oint_C \frac{f(z)}{z - \bar{\zeta}} dz \qquad (6.27)$$

Now subtracting one from the other we obtain

$$f(\zeta) = \frac{1}{2\pi i} \oint_C \frac{\zeta - \bar{\zeta}}{(z - \zeta)(z - \bar{\zeta})} f(z) dz \qquad (6.28)$$

Considering $\zeta = \xi + i\eta$ and $\bar{\zeta} = \xi - i\eta$, the above equation can be simplified as follows:

$$\begin{aligned} f(\zeta) &= \frac{1}{2\pi i} \int_{C_R} \frac{(\zeta - \bar{\zeta})}{(z - \zeta)(z - \bar{\zeta})} f(z) dz \\ &\quad + \frac{1}{\pi} \int_{-R}^{R} \frac{\eta}{(x - \xi)^2 + \eta^2} f(x, 0) dx \end{aligned} \qquad (6.29)$$

The integral along the semi-circular contour C_R goes to zero as $R \to \infty$ and hence

$$f(\xi, \eta) = \frac{1}{\pi} \int_{-\infty}^{\infty} \frac{\eta}{(x - \xi)^2 + \eta^2} f(x, 0) dx \qquad (6.30)$$

It is to be noted here that $f(x, 0)$ is the value of $f(\zeta)$ when $y = 0$ along the real x axis. Now equating the real and imaginary parts we obtain

$$\begin{aligned} u(\xi, \eta) &= \frac{1}{\pi} \int_{-\infty}^{\infty} \frac{\eta}{(x - \xi)^2 + \eta^2} u(x, 0) dx \\ v(\xi, \eta) &= \frac{1}{\pi} \int_{-\infty}^{\infty} \frac{\eta}{(x - \xi)^2 + \eta^2} v(x, 0) dx \end{aligned} \qquad (6.31)$$

Changing the roles of (x, y) with (ξ, η) we obtain

$$\begin{aligned} u(x, y) &= \frac{1}{\pi} \int_{-\infty}^{\infty} \frac{y}{(x - \xi)^2 + y^2} u(\xi, 0) d\xi \\ v(x, y) &= \frac{1}{\pi} \int_{-\infty}^{\infty} \frac{y}{(x - \xi)^2 + y^2} v(\xi, 0) d\xi \end{aligned} \qquad (6.32)$$

Now in our potential problem, we can rewrite this integral equation as

$$T(x, y) = \frac{1}{\pi} \int_{-\infty}^{\infty} \frac{y}{(x - \eta)^2 + y^2} f(\eta) d\eta \qquad (6.33)$$

where $T(x, 0) = f(x)$ for $-\infty < x < \infty$.

We shall now demonstrate applications of these theorems by practical examples.

Example 1

Show that the most general bilinear transformation which maps the upper half of the z-plane onto the interior of the unit circle in the w-plane is

$$w = e^{i\theta}\frac{z - \lambda}{z - \bar{\lambda}}$$

where λ and $\bar{\lambda}$ are two complex conjugate numbers.

Proof

The bilinear transformation is given in general by

$$w = \frac{az + b}{cz + d}$$

The following figures depict the physical situation of the problem.

By the question, the unit circle in the w-plane must be the image of the real axis in the z-plane, under the above transformation. So, we must have

$$|w| = \left|\frac{az + b}{cz + d}\right| = \frac{|a||z + b/a|}{|c||z + d/c|} = 1$$

However, from the limiting case $|z| \to \infty$, we find that

$$\frac{|a|}{|c|} = 1$$

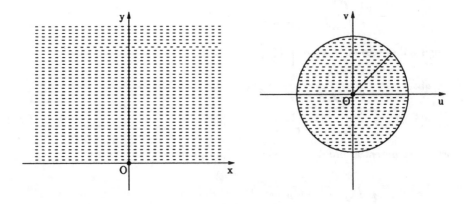

Fig. 6.11: The upper half of the z-plane to be mapped onto the interior of a unit circle in the w-plane.

This suggests that (a/c) is a complex number of absolute value 1, that means

$$\frac{a}{c} = e^{i\theta}$$

From this we conclude that for all real values of z,

$$\left| z + \frac{b}{a} \right| = \left| z + \frac{d}{c} \right|$$

or

$$\left| z - \left(-\frac{b}{a} \right) \right| = \left| z - \left(-\frac{d}{c} \right) \right|$$

The last equation expresses the fact that the complex numbers $-b/a$ and $-d/c$ are equally away from an arbitrary point z on the real axis, which is possible only if the two complex numbers are complex conjugates. Let $\lambda = -b/a$, then $\bar{\lambda} = -d/c$, and the transformation is given by

$$w = e^{i\theta} \frac{z - \lambda}{z - \bar{\lambda}}$$

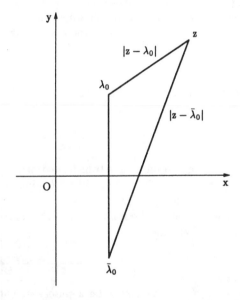

Fig. 6.12: An arbitrary point λ in the upper half of the z-plane.

So far we have enforced only the condition that the real axis in the z-plane corresponds to the circumference of the unit circle in the w-plane. It is, therefore, necessary to make sure that the upper half of the z-plane corresponds to the interior region of the unit circle in the w-plane. This can be verified by selecting a point in the plane, say, $z = \lambda$ as shown in Fig. 6.12. Clearly this point maps into the point $w = 0$, which is clearly inside the circle $|w| = 0$. Thus if λ is restricted to lie in the upper half of the z-plane, then the transformation as obtained is the required transformation.

Remark

As a special case of some interest, let us suppose that $e^{i\theta} = -1$ which simply implies that $\theta = \pi$. Let us assume that $\lambda = i$, a purely imaginary number. Then $w = -(z-i)/(z+i)$. We know that the imaginary part of w can be obtained as

$$I_m(w) = \frac{w - \bar{w}}{2i} = -\frac{1}{2i}\left[\frac{z-i}{z+i} - \frac{\bar{z}+i}{\bar{z}-i}\right]$$

After a little reduction this can be written as $I_m(w) = (z + \bar{z})/\{(z+i)(\bar{z}-i)\}$.

The denominator of the last expression is the product of a complex number with its conjugate and a positive real quantity. Thus the imaginary part of w will be positive if and only if $z + \bar{z}$ is positive. Since $z + \bar{z}$ is equal to the real part of z, this shows that the transformation $w = -(z-i)/(z+i)$ not only maps the upper half of the z-plane onto the interior of the unit circle $|w| = 1$ but does it in such a way that the first quadrant of the z-plane where $Re(z) > 0$ corresponds to the upper half of the circle where $I_m(w) > 0$ and the second quadrant of the z-plane corresponds to the lower half of the circle.

Conversely we obtain the inverse transformation as $z = -i(w-1)/(w+1)$ which maps the interior of the circle $|w| = 1$ onto the upper half of the z-plane in such a way that the upper half of the circle maps onto the first quadrant of the z-plane.

Example 2

Find a transformation which will map an infinite sector of circle $\pi/3$ onto the interior of the unit circle.

Solution

In this problem we have to use two successive transformations. Specifically, the transformation $t = z^3$ will open out the given sector in the z-plane into the upper half of the auxiliary plane t as shown in Fig. 6.13.

The upper half of the t-plane can be mapped onto the unit circle in the w-plane using the transformation in Example 1:

$$w = e^{i\theta}\frac{t - \lambda}{t - \bar{\lambda}}$$

As a special case, put $e^{i\theta} = 1$ and let λ be a purely imaginary number, say, i. Then

$$w = \frac{t - i}{t + i}$$

Thus, taking the combination of these two transformations, we obtain for the required transformation

$$w = \frac{z^3 - i}{z^3 + i}$$

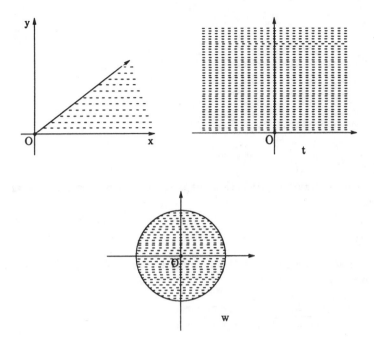

Fig. 6.13: Two successive transformations.

Example 3

Find the bilinear transformation that maps the points $z = 2, i, -2$ onto the points $w = 1, i, -1$ respectively.

Solution

Setting up the appropriate cross ratios, we have

$$\frac{(w-1)(i+1)}{(w+1)(i-1)} = \frac{(z-2)(i+2)}{(z+2)(i-2)}$$

or

$$\frac{(w-1)2i}{(w+1)2} = \frac{(z-2)(3+4i)}{(z+2)(5)}$$

$$\frac{w-1}{w+1} = \frac{(z-2)(3i-4)}{-5(z+2)}$$

or

$$w = \frac{(z-2)(3i-4) - 5(z+2)}{-5(z+2) - (z-2)(3i-4)}$$

$$= \frac{3z + 2i}{3i + 6}$$

which is the required transformation.

Example 4

Find the bilinear transformation that maps the points $z = \infty, i, 0$ onto $w = 0, i, \infty$, respectively.

Solution

The appropriate cross ratios of the 4 points in the z-plane to the 4 points in the w-plane are

$$\frac{(w - w_1)(w_2 - w_3)}{(w - w_3)(w_2 - w_1)} = \frac{(z - z_1)(z_2 - z_3)}{(z - z_3)(z_2 - z_1)}$$

Then, using the given points, we have

$$\frac{(w - 0)(i - w_3)}{(w - w_3)(i - 0)} = \frac{(z - z_1)(i - 0)}{(z - 0)(i - z_1)}$$

Now, letting $w_3 \to \infty$ and $z_1 \to \infty$, we see that

$$\frac{w(0 - 1)}{(-1)(i)} = \frac{(-1)(i)}{z(-1)}$$

or,

$$w = -\frac{1}{z}$$

which is the required transformation.

Example 5

Find a bilinear transformation which maps the upper half of the z-plane into the unit circle in the w-plane in such a way that $z = i$ is mapped into $w = 0$ while the point $z = \infty$ is mapped into $w = e^{i\theta} = -1$.

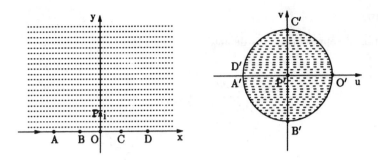

Fig. 6.14: The upper half of the z-plane is mapped into a unit circle in the w-plane.

Solution

We want the point $w = 0$ to correspond to $z = i$ and $w = -1$ to $z = \infty$. Therefore, by the transformation discussed in Example 1, we must have,

$$w = e^{i\theta}\frac{z - z_0}{z - \bar{z}_0} = (-1)\frac{z - i}{z + i} = \frac{i - z}{i + z}$$

Therefore $w = (i - z)/(i + z)$ is the required bilinear transformation. A pictorial view is given in Fig. 6.14.

The inverse transform is given by $z = i(1 - w)/(1 + w)$ which maps the interior of the unit circle in the w-plane into the upper half plane in such a way that the upper half of the circle maps onto the first quadrant of the z-plane.

Example 6

The transformation $w = e^{i\theta}(z - z_0)/(z - \bar{z}_0)$ with the condition $y \geq 0$ maps the point $z = \infty$ onto the point $w = e^{i\theta}$ which lies on the boundary of the disk $|w| = 1$. Show that if $0 < \theta < 2\pi$ and the points $z = 0, 1$ are to be mapped onto the points $w = 1, e^{i\theta}$ respectively, the bilinear transformation can be written as $w = e^{i\theta}(z + e^{-i\theta/2})/(z + e^{i\theta/2})$.

Solution

Using the bilinear transformation we can very easily obtain the relation as

$$z = \frac{(w - 1)(1 - e^{i\theta/2})}{(w - e^{i\theta/2})}(1 - e^{-i\theta/2})$$

which after simplification yields the required result.

Example 7

Find a function harmonic in the upper half of the z plane $[Im(z) > 0]$ which takes the prescribed values on the x axis given by $f(x) = 1$ for $x > 0$ and $f(x) = 0$ for $x < 0$.

Solution

The graphical representation is depicted by Fig. 6.15.

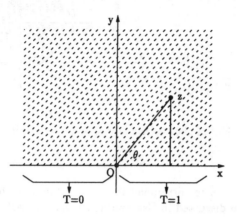

Fig. 6.15: The upper half of the z-plane.

We must solve the boundary value problem where $T(x,y)$ satisfies Laplace's equation with the boundary conditions such that

$$\frac{\partial^2 T}{\partial x^2} + \frac{\partial^2 T}{\partial y^2} = 0$$
$$T(x,0) = f(x)$$

where $f(x)$ is 1 when $x > 0$ and $f(x)$ is 0 when $x < 0$. The solution is given by

$$
\begin{aligned}
T(x,y) &= 1 + \frac{1}{\pi}[(0-1)\arg(z-0)] \\
&= 1 - \frac{1}{\pi}\arg z \\
&= 1 - \frac{1}{\pi}\arctan\left(\frac{y}{x}\right)
\end{aligned}
$$

which is the required solution.

This solution can also be obtained by Poisson's integral formula:

$$
\begin{aligned}
T(x,y) &= \frac{1}{\pi}\int_{-\infty}^{\infty}\frac{y f(\eta)d\eta}{(\eta-x)^2+y^2} \\
&= \frac{1}{\pi}\int_{0}^{\infty}\frac{y d\eta}{(\eta-x)^2+y^2} \\
&= \frac{1}{\pi}\left[\frac{\pi}{2}+\arctan\frac{x}{y}\right] \\
&= 1 - \frac{1}{\pi}\arctan\frac{y}{x}
\end{aligned}
$$

Thus it is seen that the results are identical by both the methods.

Example 8

Solve the boundary value problem

$$\frac{\partial^2 T}{\partial x^2} + \frac{\partial^2 T}{\partial y^2} = 0, \qquad y > 0$$

where the boundary conditions along the x-axis are given by

$$
\begin{aligned}
T(x,0) &= 100, & x > 1 \\
&= 0, & -1 < x < 1 \\
&= 100, & x < -1
\end{aligned}
$$

Solution

Method I:

The graphical representation of this problem has been depicted by Fig. 6.16:
By the argument principle method, we have

$$
\begin{aligned}
T(x,y) &= 100 + \frac{1}{\pi}[(0 - 100)\arg(z - 1) + (100 - 0)\arg(z + 1)] \\
&= 100 - \frac{100}{\pi}\left[\arctan\frac{y}{x - 1} - \arctan\frac{y}{x + 1}\right] \\
&= 100\left[1 - \frac{1}{\pi}\arctan\frac{2y}{x^2 + y^2 - 1}\right]
\end{aligned}
$$

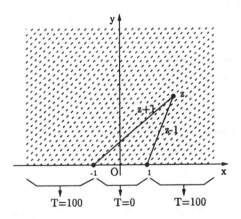

Fig. 6.16: Potential in the upper half of the z-plane.

Method II:
By the Poisson's integral method, we have

$$T(x,y) = \frac{1}{\pi}\int_{-\infty}^{\infty}\frac{yT(\eta,0)d\eta}{(\eta-x)^2+y^2}$$

$$= \frac{1}{\pi}\left[\int_{-\infty}^{-1}\frac{100y\,d\eta}{(\eta-x)^2+y^2}+\int_{1}^{\infty}\frac{100y\,d\eta}{(\eta-x)^2+y^2}\right]$$

$$= \frac{100}{\pi}\left[\arctan\frac{\eta-x}{y}\Big|_{-\infty}^{-1}+\arctan\frac{\eta-x}{y}\Big|_{1}^{\infty}\right]$$

$$= 100\left[1-\frac{1}{\pi}\arctan\frac{2y}{x^2+y^2-1}\right]$$

These two solutions are exactly identical.

Example 9

Find the steady state temperature distribution in the sheet of a metal coinciding with the first quadrant of the $z-$plane if $T = 100$ along the positive $x-$axis and $T = 0$ along the positive $y-$axis.

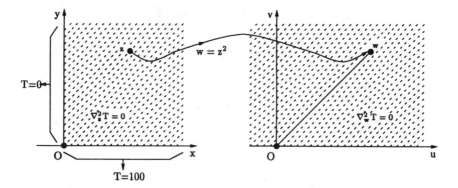

Fig. 6.17 : Conformal mapping of first quadrant in the $z-$plane onto the upper half plane of the $w-$plane.

Solution

The pictorial view of the given problem has been depicted in Fig. 6.17. The conformal mapping $w = z^2$ transforms the domain in the first quadrant in the $z-$plane to the upper half of the $w-$plane as shown in the following figure.

The solution of the problem in the $w-$plane is

$$T(u,v) = 100 + \frac{1}{\pi}[(0-100)\arg(w-0)]$$

$$= 100 \left(1 - \frac{1}{\pi} \arg w \right)$$

$$= 100 \left(1 - \frac{1}{\pi} \arctan \frac{v}{u} \right)$$

But from the conformal transformation, we have $u = x^2 - y^2$ and $v = 2xy$. Thus the solution is

$$T(x,y) = 100 \left(1 - \frac{1}{\pi} \arctan \frac{2xy}{x^2 - y^2} \right)$$

Example 10

A thin metal sheet coincides with the first quadrant of the z-plane. The upper and the lower faces of the sheet are perfectly insulated so that heat flow in the sheet is strictly two-dimensional. Find the steady state temperature at any point of the sheet if the boundary conditions are those shown in Fig. 6.18.

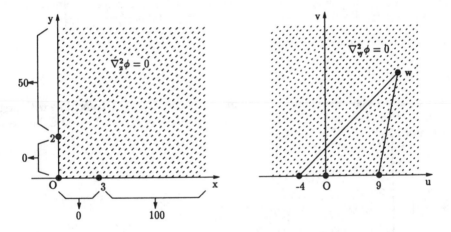

Fig. 6.18: Conformal mapping of the first quadrant in the z-plane onto the upper half plane of the w-plane.

Solution

The two-dimensional heat equation for the steady state case is Laplace's equation

$$\frac{\partial^2 \phi}{\partial x^2} + \frac{\partial^2 \phi}{\partial y^2} = 0$$

where $\phi = \phi(x,y)$ is the steady state temperature distribution in the metal sheet subject to the boundary conditions along the boundaries of the first quadrant.

To successfully obtain the required solution, we first transform the given domain in the z-plane to the upper half of the w-plane by the transformation $w = z^2$, such that $u = x^2 - y^2$ and $v = 2xy$. This reduces the problem to that of finding a solution of Laplace's equation in the upper half of the w-plane which assumes along the real axis the transformed boundary conditions.

We know that Laplace's equation in two dimensions can be satisfied by the fundamental solution in the form of logarithmic functions i.e., $\ln r$ where r is the distance between the field point and the boundary source point. But a complex function like $\ln(w - u)$ must also satisfy Laplace's equation. Thus the real and imaginary parts of $\ln(w - 9)$ and $\ln(w + 4)$ must also satisfy Laplace's equation.

Hence the required solution in the w-plane is

$$\phi(u, v) = 100 + \frac{1}{\pi}[(0 - 100)\arg(w - 9) + (50 - 0)\arg(w + 4)]$$

$$= 100 + \frac{50}{\pi}\left[-2\arctan\frac{v}{u - 9} + \arctan\frac{v}{u + 4}\right]$$

Thus transforming back to the original problem, the solution in the z-plane is

$$\phi(x, y) = 100 + \frac{50}{\pi}\left[-2\arctan\frac{2xy}{x^2 - y^2 - 9} + \arctan\frac{2xy}{x^2 - y^2 + 4}\right]$$

Hence this is the required solution.

Example 11

Find the steady state temperature distribution in a sheet of metal coinciding with first quadrant of the z-plane if $T = 100°C$ along the positive y axis, $T = 50°C$ between $x = 0$ and $x = 3$, and $T = 0°C$ to the right of $x = 3$ on the x axis.

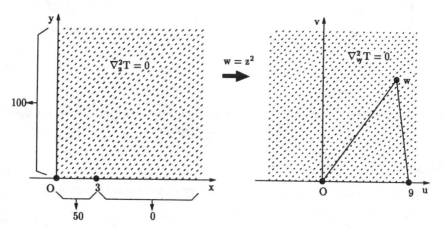

Fig. 6.19: Conformal mapping of the first quadrant in the z-plane onto the upper half-plane of the w-plane.

Solution

The pictorial view of the given problem is depicted in Fig. 6.19. By the conformal mapping $w = z^2$, i.e. $u = x^2 - y^2$ and $v = 2xy$, the entire quadrant in the z-plane can be mapped onto the upper half of the w-plane as shown above. The solution to the problem in the w-plane is

$$T(u,v) = 0 + \frac{1}{\pi}[(50 - 0)\arg(w - 9) + (100 - 50)\arg w]$$

$$= \frac{50}{\pi}\left[\arctan\frac{v}{u-9} + \arctan\frac{v}{u}\right]$$

So the solution of the problem in the z-plane is

$$T(x,y) = \frac{50}{\pi}\left[\arctan\frac{2xy}{x^2 - y^2 - 9} + \arctan\frac{2xy}{x^2 - y^2}\right]$$

Hence this is the required solution.

Exercises

1. Find the bilinear transformation that maps the points $z = 1, i, -1$ onto $w = 2, i, -2$.

2. Find the bilinear transformation that maps the points $z = -1, i, 1$ onto $w = -1, 0, i$.

3. Find the bilinear transformation that maps the points $z = 0, 1, \infty$ onto $w = \infty, i, 0$.

4. Show that a composition of two bilinear transformations is again a bilinear transformation.

5. Show that if a bilinear transformation maps z_1 into w_1 and z_2 into w_1, then either $z_1 = z_2$ or else $ad - bc = 0$. $\left[\text{Note}: w = \dfrac{az + b}{cz + d}\right]$.

6. Show that every bilinear transformation, with the exception of the identity transformation $w = z$, has at most two fixed points in the extended plane. [Hint: A fixed point of a transformation $w = f(z)$ is a point z_0 such that $z_0 = f(z_0)$.]

7. Find the fixed points of the following transformations:

   (a) $w = \dfrac{z + 1}{z - 1}$

   (b) $w = \dfrac{3z - 5}{z + 2}$

8. Show that if the origin is a fixed point of a bilinear transformation, then the transformation can be written as

$$w = \frac{az}{cz + d}$$

9. Find the most general bilinear transformation which maps the lower half of the z-plane onto the upper half of the w-plane.

10. Prove that $w = \dfrac{z}{z - 1}$ maps the upper half of the z-plane onto the upper half of the w-plane. Find the image of the circle $|z| = 1$ under this transformation.

11. Find a transformation which will map an infinite sector of angle $\pi/4$ onto the interior of the unit circle.

12. Find a transformation which will map the upper half of the unit circle onto the entire unit circle.

13. Show that if $|c| = |d|$, then the bilinear transformation, i.e. $w = \dfrac{az + b}{cz + d}$ maps the unit circle in the z-plane onto a straight line in the w-plane.

14. What is the cross ratio of the four roots of $1^{1/4}$?

15. Prove that four points, z_1, z_2, z_3, z_4 lie on a circle if and only if their cross ratio is real.

16. Use the transformation $w = \dfrac{z - 1}{z + 1}$ to find a bilinear transformation such that the disk $|z - 1| \leq 1$ is mapped onto the half plane $Re\, w \geq 0$.

17. Using the polar representation of z, show that the transformation $w = z + \dfrac{1}{z}$ maps both the upper and the lower half of the circle $|z| = 1$ onto the segment $-2 \leq u \leq 2$, $v = 0$.

18. Using the transformation in Exercise 17, show that the transformation maps the circle $|z| = a$ onto the ellipse

$$u = \left(a + \frac{1}{a} \right) \cos \theta$$

$$v = \left(a - \frac{1}{a} \right) \sin \theta$$

19. Describe the mapping $w = \cosh z$ in terms of the transformations

$$Z = e^z, \quad w = \frac{1}{2} \left(Z + \frac{1}{Z} \right)$$

20. Find the steady-state temperature distribution in the upper half of the z-plane if the portion of the x-axis between $x = -2$ and $x = 2$ is maintained at the temperature 60°C while the rest of the x-axis is maintained at 0°C.

21. Work Exercise 20 if the portion of the x-axis between $x = -2$ and $x = 2$ is maintained at 100°C, while the rest of the x-axis is maintained at 50°C.

22. Find a bilinear transformation which will map the exterior of the unit circle in the z-plane onto the upper half of the w-plane.

23. Using the result in Exercise 22, find the steady-state temperature distribution in the portion of the z-plane exterior to the unit circle if the upper half of the circumference is maintained at the temperature 50°C while the lower half is maintained at the temperature 0°C.

24. Work Exercise 23 if the area of the circle in the second quadrant is maintained at the temperature 0°C while the rest of the circumference is maintained at the temperature 50°C.

25. Find a function harmonic inside the unit circle $|z| = 1$ and taking the prescribed values given by $f(\theta) = 1$ for $0 < \theta < \pi$ and $f(\theta) = 0$ for $\pi < \theta < 2\pi$ on its circumference.

6.5 The Schwarz-Christoffel transformation

To understand the Schwarz-Christoffel transformation we need to understand the following few steps first.

Consider the following transformation by the power function

$$\xi = z^n$$

Thus it follows that the transformation

$$\xi - \xi_1 = (z - x_1)^{\theta_1/\pi}$$

$$\frac{d\xi}{dz} = \frac{\theta_1}{\pi}(z - x_1)^{(\theta_1/\pi)-1}$$

will take a segment of the x-axis containing x_1 in its interior, i.e. a straight angle with vertex at x_1, and fold it into an angle of $(\theta_1/\pi)\pi = \theta_1$ with vertex at ξ_1. Thus, if this could be done for a number of points $x_1, x_2, \cdots x_n$ on the x-axis simultaneously, the x-axis would be mapped onto a polygon whose angles are, respectively, $\theta_1, \theta_2, \theta_3, \cdots, \theta_n$. This is actually done by the following transformation.

Consider a polygon in the ξ-plane having vertices at $\xi_1, \xi_2, \cdots \xi_n$ with the corresponding interior angles $\theta_1, \theta_2, \cdots \theta_n$, respectively. Let the points $\xi_1, \xi_2, \cdots \xi_n$ map

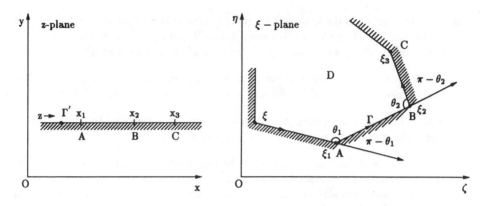

Fig. 6.20: The Schwarz-Christoffel transformation.

respectively into the points $x_1, x_2, \cdots, x_n$ on the real axis of the plane as shown in Fig. 6.20.

A transformation which maps the interior D of the polygon of the ξ-plane onto the upper half of the z-plane and the boundary of the polygon onto the real axis is given by

$$\frac{d\xi}{dz} = K(z - x_1)^{(\theta_1/\pi)-1}(z - x_2)^{(\theta_2/\pi)-1} \cdots (z - x_n)^{(\theta_n/\pi)-1} \tag{6.34}$$

To verify this, we begin with a point z on the x−axis to the left of the first of the given points $x_1, x_2, ...x_n$ and investigate the locus of its image as it moves to the right along the x−axis as shown in Fig. 6.20. From eqn (6.34) we immediately obtain the relation

$$\arg(d\xi) = \arg(K) + \left(\frac{\theta_1}{\pi} - 1\right)\arg(z - x_1) + \left(\frac{\theta_2}{\pi_-} - 1\right)\arg(z - x_2) + ...$$
$$+ \; (\frac{\theta_n}{\pi} - 1)\arg(z - x_n) + \arg(dz)$$

It is clear that until z reaches x_1, every term on the right remains constant since $z - x_1$, $z - x_2$ ..., $z - x_n$ are all negative real numbers and hence have π for their respective arguments and dz is positive and has 0 as its argument. Thus the image point ξ traces a straight line since the argument of the increment $d\xi$ remains constant. However, as z passes through x_1, the difference $z - x_1$ changes abruptly from negative to positive, and thus argument $z - x_1$ decreases abruptly from π to 0. Hence, $\arg d\xi$ changes by the amount $[(\theta_1/\pi - 1)(-\pi) = \pi - \theta_1]$. This is the precise amount through which it is necessary to turn if ξ is to begin to move in the direction of the next side of the polygon. As z moves from x_1 to x_2, the same situation exists. Thus as z traverses the x−axis, it is evident that ξ moves along the boundary of a polygon whose interior angles are precisely the given angles $\theta_1, \theta_2, ..., \theta_n$. Now the mapping function can be obtained by integrating (6.34) which gives

$$\xi = K \int (z - x_1)^{(\theta_1/\pi)-1}(z - x_2)^{(\theta_2/\pi)-1} \cdots (z - x_n)^{(\theta_n/\pi)-1} dz + L \qquad (6.35)$$

where K and L are complex constants. This formula was established by both Schwarz and Christoffel and therefore attributed to them as the Schwarz-Christoffel formula.

It is easily seen that when $z = 0$, eqn (6.35) becomes

$$\xi = \text{constant} \int_{c'} dz + L = L$$

L is an arbitrarily chosen point in the ξ-plane to correspond to $z = 0$ such that one of the vertices of the polygon can be taken to be this point. From eqn (6.34)

$$\delta \xi = K \, f(z) \delta z$$

where $f(z)$ is the integrand of (6.34). Then

$$\arg(\delta \xi) = \arg(K) + \arg f(z) + \arg(\delta z)$$

The argument of K, being a constant, merely adds a constant to the argument of $\delta \xi$ which means that the value of any K defines the orientation of the polygon. The modulus of K determines the scale of the polygon. As L, $\arg K$ and $|K|$ are arbitrary, then any three consecutive numbers, x_1, x_2, x_3 may be chosen arbitrarily. The remainder of the constants in the integrand must then be chosen to make the polygon the right shape. It is often convenient to choose one of the constants, say x_1, to be at infinity. In this case

$$K = B/(-x_1)^{(\theta_1/\pi)-1}$$

So eqn (6.35) becomes

$$\xi = B \int_{C'} \left(\frac{z - x_1}{-x_1} \right)^{(\theta_1/\pi - 1)} (z - x_2)^{(\theta_2/\pi - 1)} \cdots (z - x_n)^{(\theta_n/\pi - 1)} dz + L$$

and when $x_1 \to \infty$, this becomes

$$\xi = B \int_{C'} (z - x_2)^{(\theta_2/\pi - 1)}(z - x_3)^{(\theta_3/\pi - 1)} \cdots (z - x_n)^{(\theta_n/\pi - 1)} dz + L$$

6.6 Applications

Example 1

As an application of the Schwarz-Christoffel transformation, consider a semi-infinite rectangle in the z-plane, with two vertices at A_∞ and D_∞ at infinity where the transformation is chosen to map into the real axis in the ξ-plane such that A_∞ maps onto $\xi = -\infty$.

Solution

Consider the two other arbitrary points to map B and C onto the points $\xi = -1$ and $\xi = 1$, respectively. The vertex D_∞ obviously must map onto $\xi = +\infty$. From Fig. 6.21 it is clear that the angles of B and C are right angles $\pi/2$. Then the transformation which transforms the semi-infinite rectangle onto the real axis is given by (see Fig. 6.21)

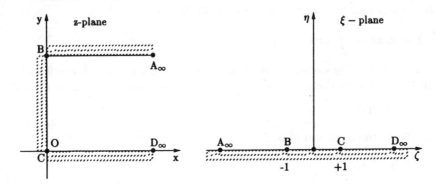

Fig. 6.21: Semi-infinite rectangle.

$$z = K \int (\xi + 1)^{-1/2}(\xi - 1)^{-1/2} d\xi + L$$

$$= K \int \frac{d\xi}{(\xi^2 - 1)^{1/2}} + L$$

On integration this yields $z = K \cosh^{-1} \xi + L$. At C, $\xi = 1$, $z = 0$, and hence $0 = K \cosh^{-1} 1 + L$. But $\cosh^{-1} 1 = 0$ and, therefore, $L = 0$. At B, $\xi = -1$, $z = ib$. Therefore, $ib = K \cosh^{-1}(-1)$. But $\cosh^{-1}(-1) = \pi i$ and hence $K = b/\pi$. Thus the required transformation is

$$z = \frac{b}{\pi} \cosh^{-1} \xi$$

which is simply

$$\xi = \cosh\left(\frac{\pi z}{b}\right)$$

This is the required solution.

Example 2

Find the transformation which maps the semi-infinite strip as shown in Fig.6.22 onto the half plane.

Solution

The interior angles at A and O are each $\pi/2$. Then by the Schwarz-Christoffel transformation we have

$$\frac{dw}{dz} = K(z-1)^{(\pi/2\pi)-1}(z+1)^{(\pi/2\pi)-1}$$
$$= K(z-1)^{-1/2}(z+1)^{-1/2}$$
$$= K(z^2-1)^{-\frac{1}{2}}$$

Integrating this equation with respect to z, we have

$$w = K\int\frac{dz}{\sqrt{z^2-1}} + L$$
$$= K\cosh^{-1}z + L$$

where K and L are two arbitrary constants. To evaluate these two constants, we use the fact that when $z = 1$, $w = 0$, and we see that $L = 0$ since $\cosh^{-1}(1) = 0$. Thus we obtain,

$$w = K\cosh^{-1}z$$

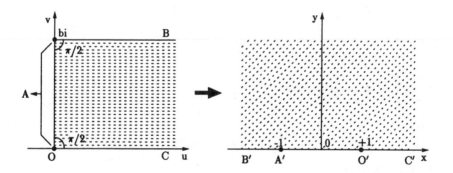

Fig. 6.22: A semi-infinite rectangular strip.

Also when $z = -1$, $w = ib$; then $K = b/\pi$ since $\cosh^{-1}(-1) = i\pi$. Hence the required transformation is

$$w = \frac{b}{\pi}\cosh^{-1}z$$

This, however, can be written as

$$z = \cosh\left(\frac{\pi w}{b}\right)$$

Note that if the coordinate of A is π, i.e. $b = \pi$, then the transformation is simply $z = \cosh w$.

Example 3

Find the transformation which maps the semi-infinite strip as shown in Fig. 6.23 below onto the half plane.

Solution

The interior angles at B and C are each $\pi/2$. Then by the Schwarz-Christoffel transformation, we obtain

$$\frac{dw}{dz} = K(z-1)^{(\pi/2\pi)-1}(z+1)^{(\pi/2\pi)-1}$$

$$= K(z-1)^{-1/2}(z+1)^{-1/2}$$

$$= K(z^2-1)^{-1/2}$$

$$= \frac{K'}{\sqrt{1-z^2}}$$

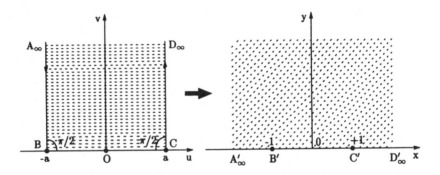

Fig. 6.23: Mapping of a semi-infinite rectangular strip.

Hence by integration

$$w = K' \arcsin z + L$$

To evaluate these constants, we notice that when $z = 1$, $w = a$; and when $z = -1$, $w = -1$. With this information, we have $L = 0$ and $K' = 2a/\pi$. Thus the required transformation in this case is

$$w = \frac{2a}{\pi} \arcsin z$$

This, however, can be written as

$$z = \sin\frac{\pi w}{2a}$$

It is to be noted that when $a = \pi/2$, the transformation simply is

$$z = \sin w$$

Example 4

Find a function which maps the interior of a triangle in the w-plane as shown in Fig. 6.24 onto the upper half of the z-plane.

Solution

By the Schwarz-Christoffel transformation, we have

$$\frac{dw}{dz} = K(z-1)^{(\gamma/\pi)-1}(z-0)^{(\alpha/\pi)-1}$$

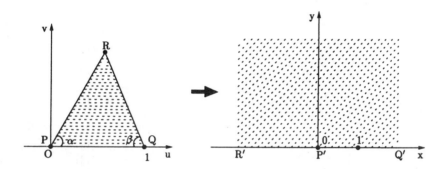

Fig. 6.24: Mapping of a triangle onto a semi-infinite rectangular strip.

Then by integration

$$w = K_1 \int_0^z t^{(\alpha/\pi)-1}(1-t)^{(\gamma/\pi)-1}dt + k_2$$

Since at $z = 0$, $w = 0$, we have $K_2 = 0$. Also since at $z = 1$, $w = 1$, we have

$$1 = K_1 \int_0^1 t^{(\alpha/\pi)-1}(1-t)^{(\gamma/\pi)-1}dt$$
$$= K_1 \beta\left(\frac{\alpha}{\pi}, \frac{\gamma}{\pi}\right)$$

where β is known as the beta function. From the properties of the beta function we know that

$$\beta\left(\frac{\alpha}{\pi}, \frac{\gamma}{\pi}\right) = \frac{\Gamma(\alpha/\pi)\Gamma(\gamma/\pi)}{\Gamma((\alpha+\gamma)/\pi)}$$

Thus the solution can be written as

$$w = \frac{1}{\beta\left((\alpha/\pi), (\gamma/\pi)\right)} \int_0^z t^{(\alpha/\pi)-1}(1-t)^{(\gamma/\pi)-1}dt$$

Example 5

Find a transformation which maps the half of the z-plane onto the infinite rectangular strip as shown in Fig. 6.25.

Solution

By the Schwarz-Christoffel transformation we have

$$\frac{dw}{dz} = K_1(z-1)^{0-1}(z+1)^{0-1}$$

$$= \frac{K_1}{(z-1)(z+1)}$$

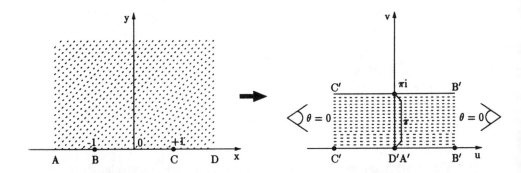

Fig. 6.25: Mapping of a half plane onto a semi-infinite rectangular strip.

Upon integration with respect to z, we have

$$w = K \ln(\frac{z-1}{z+1}) + L$$

To evaluate these constants we notice that when $z = 0$, $w = \pi i$; and when $|z|$ goes to infinity, $|w|$ becomes zero. Thus using these conditions, we obtain that $L = 0$ and $K = 1$.

Hence the required transformation is

$$w = \ln\left(\frac{z-1}{z+1}\right)$$

Example 6

Find the potential in the upper half of the z plane when the potentials along the boundary of the $x-$axis are maintained as in Fig. 6.26.

Solution

This problem can be solved by three methods:

Method I:

Solution by the Poisson formula for the half plane is as follows:

$$
\begin{aligned}
\phi(x,y) &= \frac{1}{\pi} \int_{-\infty}^{\infty} \frac{y f(\eta) d\eta}{(\eta - x)^2 + y^2} \\
&= \frac{100}{\pi} \int_{-1}^{1} \frac{y d\eta}{(\eta - x)^2 + y^2} \\
&= \frac{100}{\pi} \left[\arctan \frac{1 - x}{y} + \arctan \frac{1 + x}{y} \right] \\
&= \frac{100}{\pi} \arctan \left(\frac{2y}{x^2 + y^2 - 1} \right)
\end{aligned}
$$

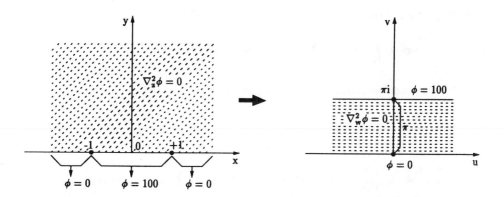

Fig. 6.26: Mapping of a half plane onto a semi-infinite rectangular strip.

Method II:

This result can be obtained by the argument principle method as follows:

$$
\begin{aligned}
\phi(x,y) &= 0 + \frac{1}{\pi}[(100 - 0)\arg(z - 1) + (0 - 100)\arg(z + 1)] \\
&= \frac{100}{\pi}[\arg(z - 1) - \arg(z + 1)] \\
&= \frac{100}{\pi} \left[\arctan \frac{y}{x - 1} - \arctan \frac{y}{x + 1} \right] \\
&= \frac{100}{\pi} \arctan \left(\frac{2y}{x^2 + y^2 - 1} \right)
\end{aligned}
$$

Method III:

The same result can be obtained by using Schwarz–Christoffel transformation as follows:

The transformation which maps the upper half of the z-plane into an infinite rectangular strip in the w-plane is $w = \ln\{(z-1)/(z+1)\}$ so that $u = \ln|(z-1)/(z+1)|$ and $v = \arg\{(z-1)/(z+1)\}$. The Laplace's equation in two dimension will take the form $d^2\phi/dv^2 = 0$. The solution of this equation is simply $\phi = K + Lv$.

Using the boundary conditions as depicted in Fig. 6.26 that means when $v = 0$, $\phi = 0$; and when $v = \pi$, $\phi = 100$ yields the following solution

$$\phi = \frac{100}{\pi}v$$

$$= \frac{100}{\pi}\arg\left(\frac{z-1}{z+1}\right)$$

$$= \frac{100}{\pi}\arctan\left(\frac{2y}{x^2+y^2-1}\right)$$

Exercises

1. Consider the transformation

$$w = A\int_{z_0}^{z}(z-x_1)^{-k_1}(z-x_2)^{-k_2}(z-x_3)^{-k_3}\,dz + B$$

In this transformation, write $B = z_0 = 0$ and $A = e^{(3\pi i)/4}$

$$x_1 = -1$$
$$x_2 = 0$$
$$x_3 = 1$$
$$k_1 = 3/4$$
$$k_2 = \frac{1}{2}$$
$$k_3 = \frac{3}{4}$$

to map the x-axis onto an isosceles right triangle. Show that the vertices of that triangle are the points

$$w_1 = bi$$
$$w_2 = 0$$
$$w_3 = b$$

where b is the positive constant

$$b = \int_0^1 (1-x^2)^{-3/4}x^{-1/2}\,dx$$

Also show that $2b = \beta\left(\frac{1}{4},\frac{1}{4}\right)$, where β is the beta function.

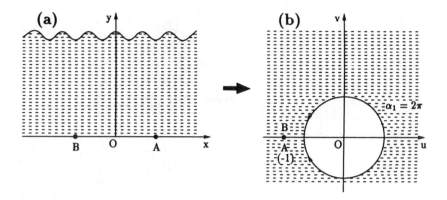

Fig. 6.27: A transformation.

2. Show that the special case

$$w = i \int_0^z (z+1)^{-1/2}(z-1)^{-1/2}z^{-1/2}dz$$

of the Schwarz-Christoffel transformation maps the x-axis onto the square with vertices, $w_1 = b_i$, $w_2 = 0$, $w_3 = b$, $w_4 = b + ib$ where the positive number b is given in terms of the beta function

$$b = \frac{1}{2}\beta\left(\frac{1}{4},\frac{1}{2}\right)$$

3. Find the transformation which will map the upper half of the z-plane onto the region in the w-plane as shown in Fig. 6.27(b).

4. Using the results of Exercise 3, find the steady-state temperature distribution in the w-plane if the upper side of the negative u-axis is kept at the temperature $0°C$ and the lower side of the negative u-axis is kept at the temperature $100°C$.

5. Using the results of Exercise 3, find the steady-state temperature distribution in the w-plane if the portion of the negative u-axis between -2 and 0 is maintained at $100°C$ while the rest of the negative u-axis is maintained at the temperature $0°C$.

6. Find the transformation which will map the exterior of the first quadrant in the w-plane onto the upper half of the z-plane as indicated in Fig. 6.28.

7. Using the results of Exercise 6, find the temperature distributions in the exterior of the first quadrant of the w-plane if the positive u-axis is kept at $0°C$ and the positive half of the v-axis is kept at $100°C$.

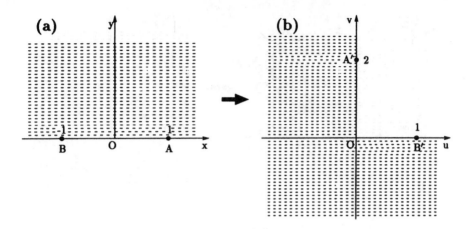

Fig. 6.28.

8. Find the transformation which will map the region shown in Fig. 6.29(b) onto the upper half of the z-plane, as indicated.

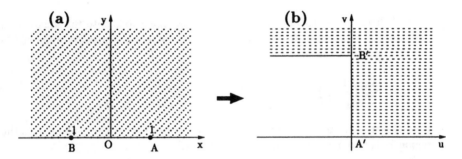

Fig. 6.29.

9. Find the transformation which will map the region shown in Fig. 6.30(b) onto the upper half of the z-plane, as indicated.

10. Find the transformation which will map the region shown in Fig. 6.31(b) onto the upper half of the z-plane, as indicated. If b goes to zero, verify that the transformation reduces to the one in Exercise 9.

11. A source is placed midway between two planes whose distance from one another is $2a$. Find the equation of the streamlines when the motion is in two dimensions; show that those particles which are at an infinite distance from the source and $\frac{1}{2}a$ from one of the boundaries, were issued from the source at an angle of $\pi/4$ with it.

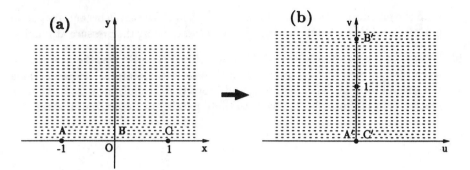

Fig. 6.30.

12. The irrotational motion in two dimensions of a fluid bounded by the lines $y = \pm b$ is due to a doublet of strength μ at the origin, the axis of the doublet being in the positive direction of the x-axis. Prove that the motion is given by

$$W = \frac{\pi \mu}{2b} \coth(\pi z / 2b)$$

Show that the points where the fluid is moving parallel to the y-axis lie on the curve

$$\cosh(\pi x / b) = \sec(\pi y / b)$$

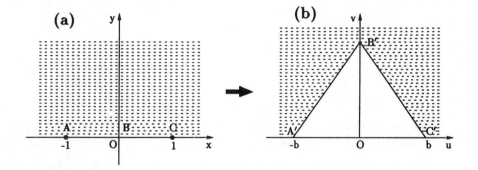

Fig. 6.31.

13. A wide stream of velocity U flows past a thin obstacle of length c which projects perpendicularly from a straight bank. Apply the Schwarz-Christoffel transformation to obtain the solution in the form

$$W^2 = U^2(z^2 + c^2)$$

Find the pressure at any point of the obstacle and show that it becomes negative if $y > c(1+k)^{1/2}(1+2k)^{1/2}$, where $k = U^2/2p_0$, p_0 being the pressure at infinity.

14. Show that the complex potential $W = m \ln \sinh(\pi z/(2a))$ gives the flow from a large vase, of breadth $2a$, through a small hole in the centre of its base. Trace the general form of the streamlines, and prove that at a distance from the base greater than its breadth the flow is sensibly parallel to the walls of the vessel.

Chapter 7

Applications to physical problems

7.1   Introduction

We shall now demonstrate the applications of the complex variables to physical problems. Specifically, we use conformal mapping to solve a number of practical problems involving Laplace's equation in two independent variables. Problems in fluid flow (hydrodynamics), heat conduction, electrostatic potential and Dirichlet problems will be treated. Applications of conformal mapping will be demonstrated to these problems from an elementary point of view.

7.2   Applications to fluid flow

7.2.1   Two dimensional flow

In this section we shall discuss two dimensional flow. It has been found experimentally that a large class of problems exists in which one of the velocity components, say w, is small when compared to the components u and v. Modelling such flows with the simplification obtained by setting $w = 0$ and allowing u and v to be functions of x and y, but not z, leads to excellent agreement between theory and observation. The flow is defined as being two dimensional.

For incompressible flow in two dimensions the continuity equation, div $\vec{V} = 0$, where $\vec{V} = (u, v, 0)$, becomes

$$\frac{\partial u}{\partial x} + \frac{\partial v}{\partial y} = 0 \tag{7.1}$$

Consider the following first order differential equation:

$$u\, dy - v\, dx = 0 \tag{7.2}$$

From the theory of first order differential equations we know that eqn (7.2) will be exact if the following condition is satisfied:

$$\frac{\partial u}{\partial x} + \frac{\partial v}{\partial y} = 0$$

which is precisely the equation of continuity (7.1). Thus there exists a function $\psi(x,y)$ such that

$$d\psi = u\,dy - v\,dx = 0 \tag{7.3}$$

Integrating eqn (7.3), we get

$$\psi(x,y) = \text{constant} \tag{7.4}$$

Here $\psi(x,y)$ is a **stream function**, since by definition, the velocity is tangential to a streamline; therefore the differential equation of streamlines can be written as

$$\frac{dy}{dx} = \frac{v}{u} \tag{7.5}$$

The relationship between the slope of a curve $\psi = $ constant and the flow velocity components in eqn (7.5) really means that such a displacement is in the direction of the local flow velocity. From eqn (7.3) we obtain

$$\frac{\partial \psi}{\partial x}dx + \frac{\partial \psi}{\partial y}dy = u\,dy - v\,dx$$

Hence

$$u = \frac{\partial \psi}{\partial y}$$

$$v = -\frac{\partial \psi}{\partial x} \tag{7.6}$$

It is noted that $\psi(x,y)$ is related to u and v; and also that this stream function exists only in two dimensional flow. For the case where the motion is **irrotational**, we must have $\text{curl }\vec{V} = 0$, that is

$$\frac{\partial v}{\partial x} - \frac{\partial u}{\partial y} = 0 \tag{7.7}$$

Equation (7.7) can be recognized as the condition for the differential equation

$$u\,dx + v\,dy = 0 \tag{7.8}$$

to be exact. Thus there exists a function $\phi(x,y)$ such that

$$d\phi = u\,dx + v\,dy \tag{7.9}$$

Hence

$$\frac{\partial \phi}{\partial x}dx + \frac{\partial \phi}{\partial y}dy = u\,dx + v\,dy$$

Therefore

$$u = \frac{\partial \phi}{\partial x}$$

$$v = \frac{\partial \phi}{\partial y} \tag{7.10}$$

and the velocity vector, $\vec{V}$ can be written as

$$\vec{V} = \vec{\text{grad}}\, \phi \tag{7.11}$$

Here $\phi(x,y)$ is called the **velocity potential**. This function exists in both two and three dimensional flow. It can be easily shown that both the stream function $\psi(x,y)$ and the velocity potential $\phi(x,y)$ satisfy Laplace's equation. Using eqn (7.6), together with the irrotational flow condition (7.7), yields

$$\frac{\partial^2 \psi}{\partial x^2} + \frac{\partial^2 \psi}{\partial y^2} = 0 \tag{7.12}$$

Similarly, using eqn (7.10), together with the continuity condition (7.1), yields

$$\frac{\partial^2 \phi}{\partial x^2} + \frac{\partial^2 \phi}{\partial y^2} = 0 \tag{7.13}$$

Remark

From the physical consideration we know that the function ψ has the property that it takes constant value along any **streamline**. The conclusion we have made with regard to eqn (7.5) can be generalized by considering any smooth curve C on which the tangential differentiation is denoted by $\partial/\partial s$ and the normal differentiation by $\partial/\partial n$ at point P on C. Then the condition that $u = \partial\phi/\partial x = \partial\psi/\partial y$ can be generalized as follows:

$$\frac{\partial \phi}{\partial n} = \frac{\partial \psi}{\partial s}$$

Now, a streamline on such a curve C on which the normal velocity $\partial\phi/\partial n$ is zero at all points along the curve and it is therefore a curve with $\partial\psi/\partial s$ is zero means the function ψ must take a constant value along the streamline. This is a very important property for a streamline.

If we now consider that the curve C is **not a streamline** but any arbitrary smooth curve, then we can obtain the difference between the values of ψ at B and A as an integral along C (see Fig. 7.1) such that

$$\psi_B - \psi_A = \int_C \frac{\partial \psi}{\partial s}\, ds$$
$$= \int_C \frac{\partial \phi}{\partial n}\, ds$$

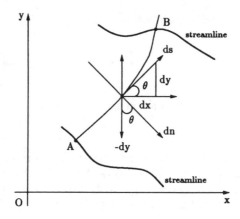

Fig. 7.1: Physical interpretation of a stream function.

Since $\partial\phi/\partial n$ is the normal velocity, the last integral represents the rate of volume flow streaming across C per unit length. For incompressible fluid, the value of the integral is independent of the path of integration and depends only on those end points A and B. Thus if we consider a closed contour Γ the integration along this closed contour the integral is definitely zero and in that situation we say the flow field is conservative. This is a physical explanation for the existence of a stream function in a two-dimensional irrotational fluid flow. Mathematically we can arrive at the same conclusion by referring to Fig. 7.1 as follows:

From this figure it is clear that

$$\frac{\partial x}{\partial s} = \cos\theta = -\frac{\partial y}{\partial n}$$

$$\frac{\partial y}{\partial s} = \sin\theta = \frac{\partial x}{\partial n}$$

Thus considering the first integral we obtain

$$
\begin{aligned}
\psi_B - \psi_A &= \int_C \frac{\partial\psi}{\partial s}\,ds \\
&= \int_C \left(\frac{\partial\psi}{\partial x}\frac{\partial x}{\partial s} + \frac{\partial\psi}{\partial y}\frac{\partial y}{\partial s}\right)\,ds \\
&= \int_C \left(\frac{\partial\psi}{\partial x}\cos\theta + \frac{\partial\psi}{\partial y}\sin\theta\right)\,ds \\
&= \int_C (-v\cos\theta + u\sin\theta)\,ds \\
&= \int_C (u\,dy - v\,dx)
\end{aligned}
$$

In a similar manner considering the second integral we obtain

$$
\begin{aligned}
\psi_B - \psi_A &= \int_C \frac{\partial \phi}{\partial n} ds \\
&= \int_C \left(\frac{\partial \phi}{\partial x} \frac{\partial x}{\partial n} + \frac{\partial \phi}{\partial y} \frac{\partial y}{\partial n} \right) ds \\
&= \int_C \left(\frac{\partial \phi}{\partial x} \sin \theta + \frac{\partial \phi}{\partial y} \cos \theta \right) ds \\
&= \int_C (u \sin \theta - v \cos \theta) ds \\
&= \int_C (u dy - v dx)
\end{aligned}
$$

These two expressions are identical and express the fact the rate of flow across all the curves joining A and B are the same. For fuller insight about this concept, the reader is referred to Lighthill (1986). In the following we shall introduce a very important entity **the complex potential** concerned with analytic function of a complex variable.

7.2.2 The complex potential

We have seen that the velocity components in two dimensional flow can be related to $\psi(x, y)$ and $\phi(x, y)$ by the following equations:

$$
\begin{aligned}
u &= \frac{\partial \psi}{\partial y} = \frac{\partial \phi}{\partial x} \\
v &= \frac{\partial \phi}{\partial y} = -\frac{\partial \psi}{\partial x}
\end{aligned} \tag{7.14}
$$

The above equations are usually defined as the Cauchy-Riemann equations. These conditions provide the necessary condition for the function

$$
W = \phi + i\psi \tag{7.15}
$$

to be an analytic function of z where $z = x + iy$. The complex function W, whose real and imaginary parts are the velocity potential and stream function, respectively, is called the Complex Potential of the flow. W is an analytic function of z and, hence

$$
\frac{dW}{dz} = \frac{\partial \phi}{\partial x} + i \frac{\partial \psi}{\partial x} = u - iv = q\, e^{-i\theta} \tag{7.16}
$$

where q = the speed of the fluid

$$
= \left| \frac{dW}{dz} \right| = \sqrt{u^2 + v^2} \tag{7.17}
$$

and θ = velocity direction relative to the real axis

$$= \tan^{-1}\frac{v}{u}$$

$$= \arg\left(\frac{dW}{dz}\right) \tag{7.18}$$

Also, at a stagnation point the fluid velocity is zero. Thus, if the complex potential, W, describing the motion is known, the stagnation points can be obtained from the equation

$$\frac{dW}{dz} = 0$$

7.2.3   Singularities in two dimensional flow

In this section some basic singularities in two dimensional potential flow are introduced. These are usually called source, sink and doublet. If the two-dimensional motion of a fluid consists of outward flow from a point, symmetrical in all directions in the reference plane, the point is called a **simple source**. A **sink** is defined as a negative source. Thus a sink is a point of inward flow at which fluid is absorbed continuously. By combining a source and a sink, a further flow singularity may be defined; this is called a **doublet**.

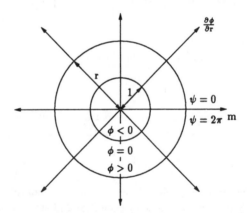

Fig. 7.2: The source.

The strength (m) of a source is defined as a measure of the flux across any curve subtending unit radius at the source. As the flow is purly radial and symmetrical, then the flux across a circle of radius r, the centre of which is the source position, can be written as

$$2\pi r \frac{\partial \phi}{\partial r}$$

The equation of continuity yields

$$2\pi r \frac{\partial \phi}{\partial r} = 2\pi m \tag{7.19}$$

The transverse velocity $(1/r)(\partial\phi/\partial\theta)$ is zero. Equation (7.19) can be written as

$$\frac{\partial \phi}{\partial r} = \frac{m}{r}$$

Integrating this equation gives

$$\phi = m \ln r \tag{7.20}$$

The Cauchy-Riemann equations in polar coordinates may be written as

$$
\begin{aligned}
v_r &= \frac{\partial \phi}{\partial r} = \frac{1}{r}\frac{\partial \psi}{\partial \theta} \\
v_\theta &= \frac{1}{r}\frac{\partial \phi}{\partial \theta} = -\frac{\partial \psi}{\partial r}
\end{aligned} \tag{7.21}
$$

Now from the first equation

$$\frac{\partial \phi}{\partial r} = \frac{m}{r} = \frac{1}{r}\frac{\partial \psi}{\partial \theta}$$

Integrating this equation gives

$$\psi = m\theta \tag{7.22}$$

Therefore the complex potential of the flow due to a source of strength m at the origin is given by

$$W = \phi + i\psi = m \ln z \tag{7.23}$$

where $z = r\,e^{i\theta}$.

Similarly the complex potential of the flow due to a sink of strength m at the origin is given by

$$W = -m \ln z \tag{7.24}$$

In Cartesian coordinates, stream function (7.22) and velocity potential (7.20) can be written as

$$
\begin{aligned}
\psi &= m \tan^{-1}\left(\frac{y}{x}\right) \\
\phi &= m \ln(x^2 + y^2)^{1/2}
\end{aligned} \tag{7.25}
$$

If the source is at the point z_0, then by a change of origin, the complex potential is

$$W = m \ln(z - z_0) \tag{7.26}$$

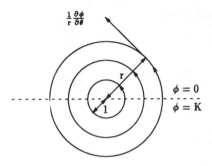

Fig. 7.3: Line vortex.

Reversing the roles of ψ and ϕ in (7.22) and (7.20) yields

$$\psi = -K \ln r$$
$$\phi = K\theta \tag{7.27}$$

By direct differentiation of (7.27), we obtain the velocity pattern

$$v_r = 0 \qquad v_\theta = \frac{K}{r} \tag{7.28}$$

Thus the flow is purely circular with the tangential velocity dropping off as $(1/r)$.

The flow pattern is shown in Fig. 7.2, there being a singularity at the origin, where the velocity is infinite; ϕ and ψ are not defined. This type of singularity in the flow field is called the **line vortex** of strength K.

By combining a source and sink in a particular way, a further flow singularity may be defined.

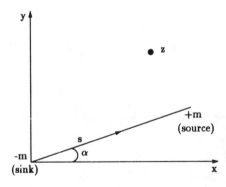

Fig. 7.4: Doublet.

Consider a source of strength m situated at a point $z_0 = r_0 e^{i\theta_0}$ and a sink of equal strength situated at the origin. The complex potential of the flow due to this system is given by

$$
\begin{aligned}
W &= m\ln(z - z_0) - m\ln z = m\ln\left(1 - \frac{z_0}{z}\right) \\
&= m\ln\left(1 - \frac{r_0 e^{i\theta_0}}{z}\right) + O\left(\frac{r_0^2}{z^2}\right) \\
&= -\frac{m r_0 e^{i\theta_0}}{z} + O\left(\frac{r_0^2}{z^2}\right)
\end{aligned}
$$

Now if $\lim_{r_0 \to 0} m r_0 = \mu$, then $W = -(\mu e^{i\theta_0})/z$.

Such a combination of source and sink is said to be a **doublet** of strength μ. Detailed discussion of flow singularities may be found in the standard textbooks, including Lamb (1945).

7.2.4   Circulation around a curve

The **circulation** around a closed contour is defined as the integral taken around the contour of the tangential component of the velocity vector $\vec{V}$.

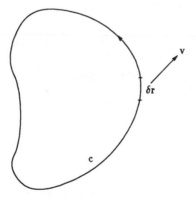

Fig. 7.5: Circulation.

Thus the circulation is

$$
\Gamma = \int_c \vec{V} \cdot d\vec{r} = \int_c (u\,dx + v\,dy + w\,dz)
$$

Using Stokes's theorem, this can be rewritten as

$$
\Gamma = \int_c \vec{V} \cdot d\vec{r} = \int\int_s \overrightarrow{\text{curl}}\vec{V} \cdot d\vec{s} = \int\int_s \vec{\omega} \cdot d\vec{s}
$$

The vector $\vec{\omega} = \text{curl}\,\vec{v}$ is called the **vorticity vector**. The rate of rotation of a fluid particle about the axes is defined as $\vec{\xi} = \frac{1}{2}\vec{\omega} = \frac{1}{2}\,\text{curl}\,\vec{V}$.

In a Cartesian coordinate system, the components of the rotation vector $\vec{\xi}$ may be written as

$$\xi_x = \frac{1}{2}\left(\frac{\partial w}{\partial y} - \frac{\partial v}{\partial z}\right)$$

$$\xi_y = \frac{1}{2}\left(\frac{\partial u}{\partial z} - \frac{\partial w}{\partial x}\right)$$

$$\xi_z = \frac{1}{2}\left(\frac{\partial v}{\partial x} - \frac{\partial u}{\partial y}\right)$$

Returning to two-dimensional fluid motion, consider the streamlines of the flow to be concentric circles. Assuming the flow is symmetrical, the transverse component of velocity will be constant for any given streamline and the radial component of velocity will be zero. Since $v_r = \partial\phi/\partial r = 0$, the circulation around a streamline of radius r will be

$$\int \vec{V} \cdot d\vec{r} = v_\theta \cdot \int_c d\vec{r} = v_\theta \cdot 2\pi r$$

$$= \left(\frac{1}{r}\frac{\partial\phi}{\partial\theta}\right) \cdot 2\pi r$$

Since the circulation around any streamline is a constant, K, regarded as positive in an anticlockwise direction, then $(1/r) \cdot (\partial\phi/\partial\theta) \cdot 2\pi r = K$, which on integration yields $\Phi = (K/2\pi)\theta$, which is the velocity potential. The Cauchy-Riemann equations in polar form are

$$v_r = \frac{1}{r}\frac{\partial\Psi}{\partial\Phi} = \frac{\partial\Phi}{\partial r}$$

$$v_\theta = \frac{\partial\Psi}{\partial r} = -\frac{1}{r}\frac{\partial\Phi}{\partial\theta}$$

Therefore $(\partial\Psi/\partial r) = -(K/2\pi)(1/r)$ which integrates to give $\Psi = -(K/2\pi)\ln r$. Thus the complex potential due to a rectilinear vortex filament is given by

$$W = \Phi + i\Psi = \frac{K}{2\pi}(\theta - i\ln r)$$

$$= -\frac{iK}{2\pi}(\ln r + i\theta) = -\frac{iK}{2\pi}\ln z$$

Using the integral $\int_c dW$, where c is a plane closed contour, a mixed flow can be produced. Writing

$$\int_C dW = \int_C \frac{dW}{dz}dz = \int_C (u - iv)(dx + i\,dy)$$

$$= \int_C (u\, dx + v\, dy) + i \int_C (u dy - v dx)$$

$$= J + i Q$$

where

$$J = \int_C u dx + v dy = \int_C \vec{V} \cdot d\vec{r}$$

the circulation in circuit C, $Q = \int_C (u dy - v dx)$ the total amount of fluid that flows outward across C. Using Green's lemma, the above integrals can be expressed as follows:

$$J = \int_C u dx + v dy = \iint_R \left(\frac{\partial v}{\partial x} - \frac{\partial u}{\partial y} \right) dx dy$$

$$Q = \int_C (u dy - v dx) = \iint_R \left(\frac{\partial u}{\partial x} + \frac{\partial v}{\partial y} \right) dx dy$$

If the integrands are analytic then the values of J and Q will be zero; conversely, if the integrands are non-analytic within the region R then J and Q are not zero. By combining one or more singularities with a uniform stream, a physical field may be obtained which will be bounded by the streamlines including the singularity from that region. The complex potential for a uniform stream may be written as $W = Uz$, where U is the uniform stream velocity in the positive direction of x axis. It is worth mentioning here that if the uniform stream originates making an angle θ with the horizontal axis, then the complex potential is given by $W = (q e^{-i\theta}) z$ where q is the speed of the fluid.

Combining this with the flow singularities such as source, sink, vortex or doublet the following physical field can be obtained.

Example 1

We can obtain the flow past a semi-infinite body by the combination of a source in a uniform stream. The complex potential may be shown to be

$$W = Uz + m \ln z$$

Then

$$\phi + i\psi = Ur\, e^{i\theta} + m \ln r + im\theta$$

Therefore

$$\phi = Ur \cos\theta + m \ln r = Ux + \frac{m}{2} \ln(x^2 + y^2)$$

$$\psi = Ur \sin\theta + m\theta = Uy + m \tan^{-1}\left(\frac{y}{x}\right)$$

The streamlines are shown in Fig. 7.6.

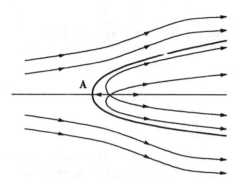

Fig. 7.6: Flow past a semi-infinite body.

Example 2

Flow past an oval shaped body may be obtained by the combination of a source and equal sink in a uniform stream.

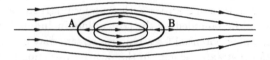

Fig. 7.7: Flow past an oval shaped body.

The complex potential is given by

$$W = Uz + m \ln(z + a) - m \ln(z - a)$$

Then

$$\phi + i\psi = Uz + m \ln\left(\frac{z+a}{z-a}\right)$$

$$= Ur\,e^{i\theta} + m \ln \frac{r_2 e^{i\theta_2}}{r_1 e^{i\theta_1}}$$

$$= Ur\,e^{i\theta} + m \ln\left(\frac{r_2}{r_1}\right) + mi(\theta_2 - \theta_1)$$

Therefore

$$\phi = Ur\,\cos\theta + m \ln\left(\frac{r_2}{r_1}\right)$$

$$\psi = Ur\,\sin\theta + m(\theta_2 - \theta_1)$$

The streamlines are shown in Fig. 7.7.

Example 3

Flow past a circular cylindrical body may be obtained by the combination of a doublet in a uniform stream. The complex potential may be shown to be

$$W = Uz + \frac{\mu}{z}$$

Then

$$\phi + i\psi = Ur\, e^{i\theta} + \frac{\mu}{r} e^{-i\theta}$$

Therefore

$$\phi = \left(Ur + \frac{\mu}{r}\right)\cos\theta$$

$$\psi = \left(Ur - \frac{\mu}{r}\right)\sin\theta$$

The streamlines are shown in Fig. 7.8.

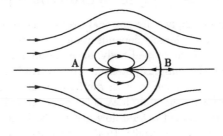

Fig. 7.8: Flow past a circular cylinder.

Example 4

Flow past a circular cylinder with circulation may be obtained by the combination of cylinder and vortex flow in a uniform stream.

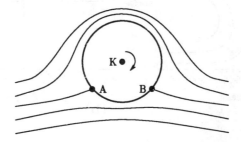

Fig. 7.9. Flow past a circular cylinder with circulation.

The complex potential may be written as

$$W = Uz + \frac{Ua^2}{z} + iK \ln z$$

Thus

$$\phi + i\psi = Ur\, e^{i\theta} + \frac{Ua^2}{r} e^{-i\theta} + iK \ln r - K\theta$$

Therefore

$$\phi = U\left(r + \frac{a^2}{r}\right)\cos\theta - K\theta$$

$$\psi = U\left(r - \frac{a^2}{r}\right)\sin\theta + K \ln r$$

The streamlines are plotted in Fig. 7.9.

7.2.5 Conformal mapping in hydrodynamics

The difficulties of investigating two-dimensional irrotational motion of an incompressible inviscid fluid about a body largely depends upon the shape of the body. As already observed the flow past a regular shaped body, for example a circular cylinder, can be obtained very easily from the complex potential. However, when the body geometry is not regular, recourse must be taken to other methods; conformal mapping techniques are useful in such situations. The main idea is to map a complicated geometry on to a simple configuration, replacing the fluid flow and the body geometry in the physical z-plane by a fictitious flow past a hypothetical configuration in the ξ-plane, where ξ is related to z by a known analytic function.

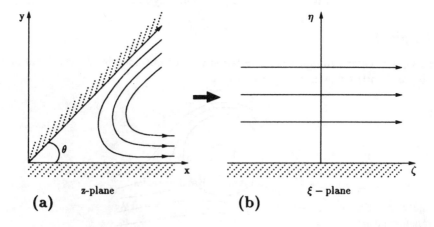

Fig. 7.10: The streamlines on the z-plane are shown as the straight lines in the ξ-plane.

As an example, let us consider the flow around a corner. We know that a corner with included angle θ in the z-plane can be mapped onto a straight line in the ξ-plane by the transformation

$$\xi = z^{\pi/\theta}$$

Figure 7.10 shows the mapping.

Another interesting example concerns the flow of an incompressible inviscid fluid past a fish-like profile with a sharp trailing edge, known as an aerofoil. By using the Joukowski transformation, this aerofoil drawn in the z-plane can be transformed into the unit circle $|\xi| = 1$ in the ξ-plane such that the flow conditions at infinity are identical.

A brief derivation of the Joukowski transformation is given below. Let the position of the cusp $z = z_0$ on the aerofoil correspond to the point $\xi = 1$ on the unit circle. Consider the transformation

$$z = f(\xi) \tag{7.29}$$

The existence of a cusp at $z = z_0$ implies a singular point in the transformation. Therefore $d\xi/dz$ must be infinite at $\xi = 1$, and in addition the flow conditions at infinity need to be identical. Thus, mathematically, we have

$$\frac{dz}{d\xi} \simeq (\xi - 1) \qquad \text{for} \quad |\xi - 1| \ll 1 \tag{7.30}$$

$$\frac{dz}{d\xi} \to 1 \qquad \text{when} \quad |\xi| \to \infty \tag{7.31}$$

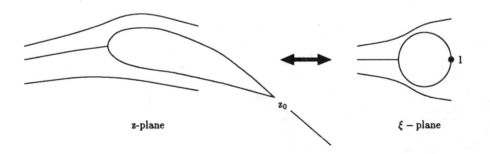

z-plane ξ - plane

Fig. 7.11: Joukowski transformation.

The analytic function $f(\xi)$ which satisfies these two conditions may be found from the transformation

$$\frac{dz}{d\xi} = \frac{(\xi - 1)(\xi - c)}{(\xi - b)^2} \tag{7.32}$$

where b and c are two arbitrary constants. Expanding the right-hand side for large ξ, yields

$$\frac{dz}{d\xi} = 1 + \frac{2b - (1+c)}{\xi} + \cdots \tag{7.33}$$

We know that, for the function to be analytic, its integral taken around a closed contour C in the ξ-plane must be zero. Thus

$$\int_C \frac{dz}{d\xi} d\xi = \int_C \frac{(\xi - 1)(\xi - c)}{(\xi - b)^2} d\xi = 0$$

provided the residue at infinity is zero, which leads to $2b = 1 + c$.

Thus the required transformation is

$$\frac{dz}{d\xi} = 1 - \frac{\frac{1}{4}(1 - c)^2}{\{\xi - \frac{1}{2}(1 + c)\}^2} \tag{7.34}$$

which after integration can be written as

$$z = \xi + \frac{\frac{1}{4}(1 - c)^2}{\xi - \frac{1}{2}(1 + c)} \tag{7.35}$$

provided $|c| < 1$ and $\frac{1}{c}(1 + c)| < 1$.

An application of the aerofoil theory can be illustrated by Blasius's theorem, which enables us to calculate the lift force and pitching moment acting on the aerofoil. Consider a cylinder of arbitrary cross-section, which is in a steady irrotational flow of fluid. The aerodynamic force per unit length of the cylinder may be reduced to a force (X, Y) and a pitching moment M, as shown in Fig. 7.12.

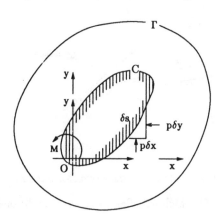

Fig. 7.12: Blasius's theorem.

Pressure forces on an element δs of the contour C may be written as δX and δY. A moment due to these forces about the origin is δM. We know $\delta X = -p\delta y$ and $\delta Y = p\delta x$ then $\delta M = x\delta Y - y\delta X = p(x\delta x + y\delta y)$.

Thus

$$
\begin{aligned}
\delta X - i\delta Y &= -p\delta y - ip\delta x \\
&= -ip(\delta x - i\delta y) \\
&= -ip\delta\bar{z}
\end{aligned}
$$

and therefore $\delta(X - iY) = -ip\delta\bar{z}$. Integrating along the contour C gives

$$
X - iY = -i\int_C p d\bar{z}
$$

which, on using Bernoulli's equation, becomes

$$
-i\int_C \left[\text{constant} - \frac{1}{2}\rho q^2\right] d\bar{z}
$$

Thus

$$
X - iY = -i\int_C (\text{constant})d\bar{z} + \frac{1}{2}i\rho\int_C q^2 d\bar{z}
$$

$$
\int_C d\bar{z} = 0, \quad \text{and} \quad q^2 = \frac{dW}{dz}\cdot\frac{d\bar{W}}{d\bar{z}}
$$

where $\bar{q} = dW/dz = u - iv$ and $q^2 = u^2 + v^2$. Since W is the complex potential, hence

$$
X - iY = i\frac{\rho}{2}\int_C \frac{dW}{dz}d\bar{W}
$$

Since on C the stream function ψ is a constant, then $dW = d\bar{W}$. The above equation may be expressed as

$$
\begin{aligned}
X - iY &= i\frac{\rho}{2}\int_C \left(\frac{dW}{dz}\right)^2 dz \\
&= i\frac{\rho}{2}\int_\Gamma \left(\frac{dW}{dz}\right)^2 dz
\end{aligned}
\tag{7.36}
$$

where Γ is a contour reconcilable with C.

To find the moment about the origin, note that

$$
\delta M = p(x\delta x + y\delta y)
$$

Since $Re(z\delta\bar{z}) = x\delta x + y\delta y$, we have

$$
\begin{aligned}
\delta M &= Re\,[pz\delta\bar{z}] \\
&= Re\left[\left(\text{constant} - \frac{1}{2}\rho q^2\right)z\delta\bar{z}\right]
\end{aligned}
$$

Integrating along the contour C

$$M = Re \int_C \left[\text{constant} - \frac{1}{2}\rho q^2 \right] z\delta\bar{z}$$

$$= Re \left[-\frac{1}{2}\rho \int_C q^2 z d\bar{z} \right]$$

because the integration along C, i.e. $Re \int_C z d\bar{z} = 0$. Thus we obtain

$$M = Re \left[-\frac{1}{2}\rho \int_C z (\frac{dW}{dz})^2 dz \right] \tag{7.37}$$

$$= Re \left[-\frac{\rho}{2} \int_\Gamma z \left(\frac{dW}{dz} \right)^2 dz \right] \tag{7.38}$$

Equations (7.36) and (7.37) are known as **Blasius's theorem**.

7.2.6   Lift force

Let dW/dz have no singularities outside the contour C. Then we may assume Γ to be a contour at infinity. Thus, it is reasonable to expand dW/dz in inverse powers of z as follows:

$$\frac{dW}{dz} = a_0 + \frac{a_1}{z} + \frac{a_2}{z^2} + \cdots$$

It is known that the fluid velocity must tend to a constant which is the velocity of the uniform stream.

Suppose the uniform stream velocity is U at an angle α to the real axis. Then

$$|z| \to \infty \qquad \frac{dW}{dz} \to Ue^{-i\alpha} = a_0$$

where α is the incidence angle or angle of attack of the aerofoil. Thus

$$W = Ue^{-i\alpha}z + a_1 \ln z - \frac{a_2}{z} + \cdots$$

The second term, being the only multiple valued function, must represent the circulation $2\pi K$ around the aerofoil. Hence $a_1 = iK$.

Thus

$$\frac{dW}{dz} = Ue^{-i\alpha} + \frac{iK}{z} + \frac{a_2}{z^2} + \cdots$$

$$\left(\frac{dW}{dz} \right)^2 = U^2 e^{-2i\alpha} + (2Ue^{-i\alpha}) \cdot \frac{iK}{z} + O\left(\frac{1}{z^2} \right)$$

We know from the theory of residue that the coefficient of $(1/z)$ which is precisely $2Ue^{-i\alpha}(iK)$ is the residue of the given function at $z = 0$.

Therefore

$$X - iY = \int_\Gamma \frac{i}{2}\rho \left(\frac{dW}{dz}\right)^2 dz$$

$$= \left(\frac{i}{2}\rho\right)(2\pi i) \times \text{residue}$$

$$= \left(\frac{i}{2}\rho\right)(2\pi i)(2Ue^{-i\alpha} \cdot iK)$$

$$X - iY = -2\pi i\rho Ue^{-i\alpha}K = -2\pi\rho UK[i\cos\alpha + \sin\alpha]$$

On comparing real and imaginary parts, $X = -2\pi\rho UK\sin\alpha$ and $Y = 2\pi\rho UK\cos\alpha$. The lift force is given by

$$L = \sqrt{X^2 + Y^2}$$

$$= 2\pi\rho UK$$

$$= (2\pi) \times (\text{density}) \times (\text{mainstream}) \times (\text{circulation})$$

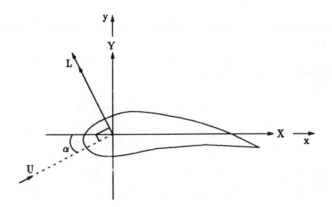

Fig. 7.13: Lift force.

This lift force is perpendicular to the mainstream U and is independent of the form of the aerofoil. This result is attributed to Kutta and Joukowski.

7.2.7   Applications

Example 5

The complex potential of a fluid flow is given by

$$W(z) = V_0 \left(z + \frac{a^2}{z} \right)$$

where V_0 and a are positive constants.

(a) Obtain equations for streamlines and equipotential lines, represent them graphically and interpret them physically.

(b) Show that we can interpret the flow as that around a circular obstacle of radius a.

(c) Find the velocity at any point and determine its value from the obstacle.

(d) Find the stagnation points.

Solution

The complex potential is given by

(a) $$W = V_0 \left(z + \frac{a^2}{z} \right)$$

Therefore

$$
\begin{aligned}
\phi + i\psi &= V_0 \left(re^{i\theta} + \frac{a^2}{r} e^{-i\theta} \right) \\
&= V_0 \left[r(\cos\theta + i\sin\theta) + \frac{a^2}{r}(\cos\theta - i\sin\theta) \right] \\
&= V_0 \left[\left(r + \frac{a^2}{r} \right) \cos\theta + i \left(r - \frac{a^2}{r} \right) \sin\theta \right]
\end{aligned}
$$

Thus, equating real and imaginary parts $\phi = V_0 (r + (a^2/r)) \cos\theta$ and $\psi = V_0 (r - (a^2/r)) \sin\theta$ the streamlines are given by $\psi = constant = \beta$.
Therefore $V_0(r - \frac{a^2}{r}) \sin\theta = \beta$.

The streamlines are indicated by the heavy curves and show the actual paths taken by fluid particles. Note that $\psi = 0$ corresponds to $r = a$ and $\theta = 0$ or π. The equipotential lines are given by $\phi = constant = \alpha$ or $V_0 (r + (a^2/r)) \cos\theta = \alpha$. These are indicated by the ϕ curves and are orthogonal to the family of streamlines, i.e. the ψ curves.

(b) The circle $r = a$, which represents a streamline, and since there can be no flow across a streamline, it can be considered as a circular obstacle of radius a placed in the path of the fluid.

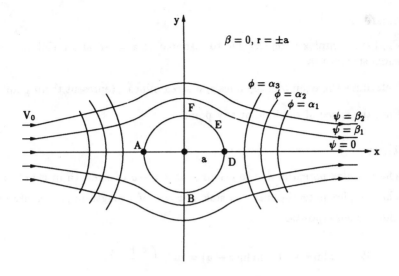

$\beta = 0, r = \pm a$

$\phi = \alpha_3$
$\phi = \alpha_2$
$\phi = \alpha_1$

$\psi = \beta_2$
$\psi = \beta_1$
$\psi = 0$

Fig. 7.14: Streamlines and equipotential lines.

(c) We have

$$\frac{dW}{dz} = W' = V_0 \left(1 - \frac{a^2}{z^2}\right) = V_0 \left(1 - \frac{a^2}{r^2}e^{-2i\theta}\right)$$

Therefore

$$u - iv = V_0 \left(1 - \frac{a^2}{r^2}\cos 2\theta\right) + i\frac{V_0 a^2}{r^2}\sin 2\theta$$

Equating real and imaginary parts we have $u = V_0 (1 - (a^2/r^2)\cos 2\theta)$ and $v = -V_0 ((a^2/r^2)\sin 2\theta)$

The absolute velocity is given by

$$q = V_0 \sqrt{\left\{1 - \frac{a^2}{r^2}\cos 2\theta\right\}^2 + \frac{a^4}{r^4}\sin^2 2\theta} = V_0 \sqrt{1 - \frac{2a^2\cos 2\theta}{r^2} + \frac{a^4}{r^4}}$$

Far from the obstacle, as $r \to \infty$, we see that $q = V_0$ approximately, that is the fluid is travelling in the direction of the positive x with constant speed V_0.

(d) The stagnation points are points at which the velocity is zero, that is $(dW/dz) = 0$ and so we have $z = a$ and $z = -a$. The stagnation points are therefore A and D.

Example 6

(a) Find the complex potential due to a source at $z = -a$ and a sink at $z = a$ of equal strength m.

(b) Determine the equipotential lines and streamlines. Represent them graphically.

(c) Find the speed of the fluid at any point.

Solution

(a) The complex potential due to a source at $z = -a$ of strength m is $m\ln(z + a)$.
The complex potential due to a sink at $z = a$ of strength m is $-m\ln(z - a)$.
Then by superposition

$$W = m\ln(z + a) - m\ln(z - a) = m\ln\left(\frac{z+a}{z-a}\right)$$

(b) Let us suppose that $z + a = r_1 e^{i\theta_1}$ and $z - a = r_2 e^{i\theta_2}$. Then

$$W = \phi + i\psi = m\ln\frac{r_1 e^{i\theta_1}}{r_2 e^{i\theta_2}} = m\ln\left\{\frac{r_1}{r_2}e^{i(\theta_1-\theta_2)}\right\} = m\ln\frac{r_1}{r_2} + mi(\theta_1 - \theta_2)$$

On identifying real and imaginary parts, we have $\phi = m\ln(r_1/r_2)$ and $\psi = m(\theta_1 - \theta_2)$.

Thus the equipotential lines and the streamlines are, respectively, given by

$$\phi = m\ln\left(\frac{r_1}{r_2}\right) = \alpha$$

$$\psi = m(\theta_1 - \theta_2) = \beta$$

Using $r_1 = \sqrt{(x + a)^2 + y^2}$ and $r_2 = \sqrt{(x - a)^2 + y^2}$, we obtain
$\theta_1 = \tan^{-1}\{y/(x + a)\}$ and $\theta_2 = \tan^{-1}\{y/(x - a)\}$.
The equipotential lines are given by

$$\frac{r_1}{r_2} = e^{\alpha/m}$$

which on substitution of the above expression gives

$$\frac{\sqrt{(x + a)^2 + y^2}}{\sqrt{(x - a)^2 + y^2}} = e^{\alpha/m} = \lambda$$

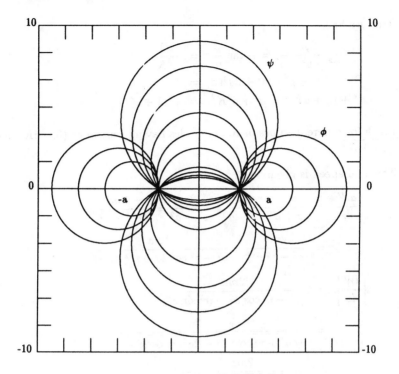

Fig. 7.15: Streamlines and equipotential lines.

Thus we have

$$(x+a)^2 + y^2 = \lambda^2\{(x-a)^2 + y^2\}$$
$$x^2 + 2ax + y^2 + a^2 = \lambda^2\{x^2 - 2ax + a^2 + y^2\}$$
$$x^2 + a^2 + y^2 + 2ax\frac{1+\lambda^2}{1-\lambda^2} = 0, \quad \lambda \neq \pm 1$$
$$\left\{x - a\frac{\lambda^2+1}{\lambda^2-1}\right\}^2 + y^2 = \frac{4a^2\lambda^2}{(\lambda^2-1)^2}$$

which represents circles having centres at $(a\{(\lambda^2+1)/(\lambda^2-1)\},0)$ and radii equal to $2a\lambda/(\lambda^2-1)$. When $\lambda = \pm 1$, the equipotential line is given by $x = 0$.

Streamlines are given by

$$\tan^{-1}\left[\frac{y}{x+a} - \tan^{-1}\frac{y}{x-a}\right] = \frac{\beta}{m}$$

and therefore

$$\tan^{-1}\left[\frac{y/(x+a) - y/(x-a)}{1 + y^2/(x^2-a^2)}\right] = \left(\frac{\beta}{m}\right)$$

Thus we have

$$\frac{2ya}{x^2 + y^2 - a^2} = -\tan\frac{\beta}{m} = -\frac{1}{K}$$

$$x^2 + y^2 - a^2 + 2yKa = 0$$

$$x^2 + (y + aK)^2 = a^2 + a^2K^2 = a^2(1 + K^2)$$

which for different values of K are circles having centres at $(0, -aK)$ and radii $a\sqrt{1 + K^2}$.

(c) The fluid speed is given by

$$\frac{dW}{dz} = \frac{m}{z + a} - \frac{m}{z - a} = \frac{-2ma}{z^2 - a^2} = \frac{-2ma}{2e^{2i\theta} - a^2}$$

$$= \frac{2ma}{a^2 - r^2 e^{2i\theta}} = \frac{2ma}{a^2 - r^2\cos 2\theta - ir^2\sin 2\theta}$$

$$\left|\frac{dW}{dz}\right| = \frac{2ma}{|a^2 - r^2\cos 2\theta - ir^2\sin 2\theta|}$$

$$= \frac{2ma}{\sqrt{a^4 + r^4\cos^2 2\theta - 2a^2r^2\cos 2\theta + r^4\sin^2 2\theta}}$$

$$= \frac{2ma}{\sqrt{a^4 + r^4 - 2a^2r^2\cos 2\theta}}$$

Example 7

Two sources each of strength m are placed at points $(-a, 0)$ and $(a, 0)$ and a sink of strength $2m$ is placed at the origin. Show that the streamlines are the curves

$$(x^2 + y^2)^2 = a^2(x^2 - y^2 + \lambda xy)$$

where λ is a variable parameter.

Solution

The complex potential due to a source at $z = -a$ of strength m is $m\ln(z + a)$.
The complex potential due to a source at $z = a$ of strength m is $m\ln(z - a)$.
The complex potential due to a sink at $z = 0$ of strength $2m$ is $-2m\ln z$.
Therefore by superposition, the total complex potential is

$$W = m\ln(z + a) + m\ln(z - a) - 2m\ln(z) = m\ln\left\{\frac{(z + a)(z - a)}{z^2}\right\}$$

Now let us define that $z + a = r_1 e^{i\theta_1}$, $z - a = r_2 e^{i\theta_2}$ and $z = re^{i\theta}$,

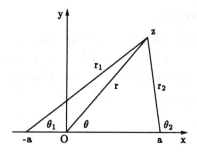

Fig. 7.16: Complex potential.

where

$$r_1 = \sqrt{(x+a)^2 + y^2}$$
$$r_2 = \sqrt{(x-a)^2 + y^2}$$
$$r = \sqrt{x^2 + y^2}$$

and

$$\theta_1 = \tan^{-1} \frac{y}{x+a}$$

$$\theta_2 = \tan^{-1} \frac{y}{x-a}$$

$$\theta = \tan^{-1} \frac{y}{x}$$

Thus we have

$$W = \phi + i\psi = m \ln \left\{ \frac{r_1 r_2}{r^2} e^{i(\theta_1 + \theta_2 - 2\theta)} \right\}$$
$$= m \ln \left(\frac{r_1 r_2}{r^2} \right) + im(\theta_1 + \theta_2 - 2\theta)$$

Therefore $\phi = m \ln (r_1 r_2 / r^2)$ is the velocity potential, and $\psi = m \ln\{\theta_1 + \theta_2 - 2\theta\}$ is the stream function.

From the above expressions for θ_1 and θ_2 we have

$$\theta_1 + \theta_2 - 2\theta = \tan^{-1} \frac{y}{x+a} + \tan^{-1} \frac{y}{x-a} - 2\tan^{-1} \frac{y}{x}$$

$$\theta_1 + \theta_2 - 2\theta = \tan^{-1} \frac{2xy/(x^2 - a^2)}{1 - y^2/(x^2 - a^2)} - \tan^{-1} \frac{2y/x}{1 - y^2/x^2}$$

$$= \tan^{-1} \frac{2xy}{x^2 - y^2 - a^2} - \tan^{-1} \frac{2xy}{x^2 - y^2}$$

$$= \tan^{-1} \frac{2xya^2}{(x^2 - y^2)(x^2 - y^2 - a^2) + 4x^2y^2}$$

The streamlines are given by

$$(x^2 - y^2)(x^2 - y^2 - a^2) + 4x^2y^2 = 2a^2xy(\text{constant})$$
$$(x^2 - y^2)^2 + 4x^2y^2 = a^2(x^2 - y^2 + \lambda xy)$$
$$(x^2 + y^2)^2 = a^2(x^2 - y^2 + \lambda xy)$$

where λ is a variable parameter.

Example 8

Prove that the pressure distribution P, around a circular cylinder fixed with its axis orthogonal to a uniform stream, V_0, is given by

$$p = \frac{1}{2}\rho V_0^2(1 - 4\sin^2\theta)$$

Solution

The complex velocity potential is

$$W(z) = V_0\left(z + \frac{a^2}{z}\right)$$

Therefore

$$
\begin{aligned}
\phi + i\psi &= V_0\left(re^{i\theta} + \frac{a^2}{r}e^{-i\theta}\right) \\
&= V_0\left[\left(r\cos\theta + \frac{a^2}{r}\cos\theta\right) + i\left(r\sin\theta - \frac{a^2}{r}\sin\theta\right)\right]
\end{aligned}
$$

Thus $\phi = V_0(r + a^2/r)\cos\theta$ and $\psi = V_0(r - a^2/r)\sin\theta$.

By Cauchy-Riemann conditions $V_r = \partial\phi/\partial r = (1/r)(\partial\psi/\partial\theta)$ and $V_\theta = (1/r)(\partial\phi/\partial\theta) = -\partial\psi/\partial r$, where V_r is the radial velocity component and V_θ is the tangential velocity component.

But, on the surface of the cylinder, $V_r = 0$.

Therefore

$$V_\theta = \left(\frac{1}{r}\frac{\partial\phi}{\partial\theta}\right)_{r=a} = -2V_0\sin\theta$$

From Bernoulli's equation

$$p + \rho\frac{\partial\phi}{\partial t} + \frac{\rho}{2}q^2 = C$$

we obtain for steady flows ($\partial\phi/\partial t = 0$) that $p + (\rho/2)q^2 = C$. Also we know that, at infinity, when $r \to \infty$, $q = V_0$ and $p = 0$. Therefore, $C = (\rho/2)V_0^2$. Thus, we have $p = (\rho/2)(V_0^2 - q^2)$. However, $q^2 = V_\theta^2$ at $r = a$ such that $q^2 = 4V_0^2\sin^2\theta$ and hence $p = \frac{\rho}{2}V_0^2(1 - 4\sin^2\theta)$.

Exercises

1. Find the streamlines and the equipotential lines for the fluid motion in which the complex potential is given by

   (a) $W = z + 2$

   (b) $W = z^2 + 4z$

   (c) $W = 3^{-z}$

   (d) $W = \sin z$

2. Determine the streamlines and the equipotential lines for the fluid flow whose complex potential is given by

   $$W = U_0 \left(z + \frac{a^2}{z} \right)$$

   Sketch these lines and confirm that they are orthogonal.

3. Discuss the fluid flow due to two sources at $z = \pm a$ and a sink at $z = 0$ where the strengths of sources and sink have the same value of m.

4. Sketch the streamlines and the equipotential lines for fluid motion in which the complex potential is given by

   $$W = U_0 z + m \ln z$$

   where U_0 is the uniform speed of the fluid and m is the source strength.

5. Find the complex potential corresponding to a sink at $z = -a$ and a source at $z = a$ when the strengths all have equal magnitude. Find the velocity potential and the stream function and sketch the equipotential curves and streamlines.

6. Discuss the fluid flow due to three sources of equal strength m located at $z = -1, 0, 1$.

7. Determine the velocity potential and the stream function when the complex potential is $W = z^2$. Find the velocity of the fluid at the points

   (a) $z = 2 + i$

   (b) $z = -2 + i$

8. In Exercise 7, find the locus of points where the absolute value of the velocity to speed of the fluid is a constant.

9. Find the velocity potential and the stream function for the flow whose complex potential is $W = \cos^{-1} z$. Sketch the flow pattern.

10. Find the velocity potential and the stream function for the flow whose complex potential is $W = \cosh^{-1} z$. Sketch the flow pattern.

11. Find the complex potential of the flow consisting of two vortices of equal strength but of opposite sign placed at $z = -a$ and $z = a$. Determine the velocity potential and stream function. Sketch the equipotential lines and streamlines.

12. Determine the potential at any point in the portion of the first quadrant between the hyperbolas $xy = 1$ and $xy = 5$ if the first hyperbola is maintained at the potential 10 and the second is maintained at 100.

13. Find the potential at any point between the annulus of the concentric circles $|z| = 2$ and $|z| = 4$ if the first circle is maintained at the potential 100 and the second at the potential 50.

14. Consider a circle with its centre at $z = a(0 < a < 1)$ on the x-axis which passes through the point $z = -1$. This circle is subjected to the transformation $w = z + 1/z$. Individual non-zero points $z = e^{i\theta}$ can be mapped geometrically by adding the vector $1/(r\,e^{i\theta})$ to the vector z. Indicate by mapping some points that the image of the circle is a profile of a special Joukowski airfoil.

15. Show that the inverse of the transformation $W = z + 1/z$ is given, with z and W interchanged. Find the complex potential for the flow around the airfoil when the velocity V of the fluid at an infinite distance from the origin is a real constant A.

16. Find the complex potential for the flow external to the cylinder $|z| = a$ due to a source of strength m at $z = 2a$.

    (a) Prove that the force on the cylinder is $(\rho m^2)/(12\pi a)$; find its direction.

    (b) Prove that $\int_c (dW/dz)^2 dz = 0$, where C is the contour $|z| = 3a$.

    (c) Hence show that, in this flow, the equal and opposite force which acts on the source is given by Blasius's expression evaluated for a contour which encloses the source alone.

17. Prove that the transformation $\xi = e^{\pi z/2a}$ transforms a strip of breadth $2a$ into a half plane. Consider the case where incompressible fluid fills the space between two infinite parallel planes at a distance $2a$ apart. Two line sources of strengths $+m$ and $-m$ respectively, are parallel to each other and lie midway between the two planes a distance $2b$ apart. Show that the complex potential function is

$$W = m \ln \left\{ \frac{\exp(\pi z/a) + \exp(\pi b/a)}{\exp(\pi z/a) + \exp(-\pi b/a)} \right\}$$

18. Show that $W = K \ln(z^2 - a^2)$ gives the motion due to a two-dimensional source in the presence of a fixed wall, and, by using the transformation given by

$$\frac{d\xi}{dz} = \frac{B}{(z^2 - 1)^{1/2}}$$

obtain the solution for such a source in a semi-infinite rectangle.

19. A source is placed midway between two planes whose distance from one another is $2a$. Find the equation of the streamlines when the motion is in two dimensions; show that those particles which are at an infinite distance from the source and $\frac{1}{2}a$ from one of the boundaries, were issued from the source at an angle of $\pi/4$ with it.

20. The irrotational motion in two dimensions of a fluid bounded by the lines $y = \pm b$ is due to a doublet of strength μ at the origin, the axis of the doublet being in the positive direction of the x-axis. Prove that the motion is given by

$$W = \frac{\pi \mu}{2b} \coth(\pi z/2b)$$

Show that the points where the fluid is moving parallel to the y-axis lie on the curve

$$\cosh(\pi x/b) = \sec(\pi y/b)$$

21. A wide stream of velocity U flows past a thin obstacle of length c which projects perpendicularly from a straight bank. Apply the Schwarz-Christoffel transformation to obtain the solution in the form

$$W^2 = U^2(z^2 + c^2)$$

22. Show that the complex potential $W = m \ln \sinh(\pi z/2a)$ gives the flow from a large vase, of breadth $2a$, through a small hole in the centre of its base. Trace the general form of the streamlines, and prove that at a distance from the base greater than its breadth the flow is sensibly parallel to the walls of the vessel.

7.3 Applications to heat flow

7.3.1 Heat flux in two dimension

Consider a solid which has a varying temperature distribution. In the process of heat conduction, the quantity of heat conducted per unit time per unit area across a surface located in the solid is called the heat flux and is mathematically defined by Fourier's Law

$$Q = -k \operatorname{grad} \phi \tag{7.39}$$

where Q is the heat flux, ϕ is the temperature and k is known as the thermal conductivity of the material of the solid.

If we restrict ourselves to a two dimensional heat flow problem, then

$$Q = -k\left(\frac{\partial \phi}{\partial x} + i\frac{\partial \phi}{\partial y}\right)$$

$$= Q_x + iQ_y \tag{7.40}$$

where $Q_x = -k\partial\phi/\partial x$ and $Q_y = -k\partial\phi/\partial y$.

Let C be any simple closed curve in the z−plane which represents the cross section of a cylinder. If Q_n and Q_t are the normal and tangential components of the heat flux vector Q and if we consider the steady heat flow such that there is no net accumulation of heat inside C then we have (assuming no sources or sink inside C)

$$\oint_C Q_n ds = \oint_C Q_x dy - Q_y dx = 0 \tag{7.41}$$

$$\oint_c Q_t ds = \oint_c Q_x dx + Q_y dy = 0 \tag{7.42}$$

Using Green's lemma, eqn (7.41) can be reduced to

$$\frac{\partial Q_x}{\partial x} + \frac{\partial Q_y}{\partial y} = 0 \tag{7.43}$$

Using relation (7.41) yields

$$\frac{\partial^2 \phi}{\partial x^2} + \frac{\partial^2 \phi}{\partial y^2} = 0 \tag{7.44}$$

which means that ϕ is harmonic. Introducing the harmonic conjugate function ψ, we see that

$$W(z) = \phi(x,y) + i\Psi(x,y) \tag{7.45}$$

is analytic. The families of curves

$$\left.\begin{array}{l} \phi(x,y) = c_1 \\ \psi(x,y) = c_2 \end{array}\right\} \tag{7.46}$$

are respectively called isothermal lines and flux lines (streamline), while $W(z)$ is called the complex temperature distribution.

Example 1

A semi-infinite slab (see Fig. 7.17(a)) has its boundaries maintained at the indicated temperatures where T is a constant. Find the steady state temperature.

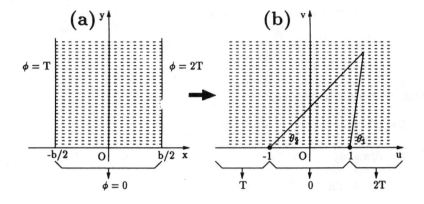

Fig. 7.17.

Solution

Method I:

The shaded area of the z-plane shown in Fig. 7.17(a) can be mapped into the upper half of the w-plane (Fig. 7.17(b)) by the mapping function

$$w = \sin\left(\frac{\pi z}{b}\right)$$

This is equivalent to the following two functions

$$u = \sin\left(\frac{\pi x}{b}\right)\cosh\left(\pi\frac{y}{b}\right)$$

$$v = \cos\left(\frac{\pi x}{b}\right)\sinh\left(\pi\frac{y}{b}\right)$$

We must now solve the equivalent problem in the w plane. This problem is a Dirichlet problem for the upper half of the plane w.

The function, $\phi = A\theta_1 + B\theta_2 + C = A\arg(w-1) + B\arg(w+1) + C$ where A, B and C are real constants, is harmonic since it is the real part of the complex potential $W = \phi + i\psi$ given by

$$W = C - i(A\ln(w-1) + B\ln(w+1))$$

To determine A, B and C note that the boundary conditions are:

$$\phi = 2T \qquad \text{for } u > 1 \qquad\qquad \text{i.e. } \theta_1 = \theta_2 = 0$$
$$\phi = T \qquad \text{for } u < -1 \qquad\qquad \text{i.e. } \theta_1 = \pi = \theta_2$$
$$\phi = 0 \qquad \text{for } -1 < u < 1 \qquad \text{i.e. } \theta_1 = \pi, \theta_2 = 0$$

Thus, using these boundary conditions, we obtain

- $2T = A.0 + B.0 + C$

- $T = A\pi + B\pi + C$

- $0 = A.\pi + 0.B + C$

Hence we have, after solving,

$$\left.\begin{array}{l} C = 2T \\ A = -2T/\pi \\ B = T/\pi \end{array}\right\}$$

The solution in the w plane is then

$$\phi = -\frac{2T}{\pi}\tan^{-1}\left(\frac{v}{u-1}\right) + \frac{T}{\pi}\tan^{-1}\left(\frac{v}{u+1}\right) + 2T$$

and the required solution to the problem in the z-plane is therefore,

$$\begin{aligned} \phi \;=\; & \frac{T}{\pi}\tan^{-1}\left\{\frac{\cos(\pi x/b)\sinh(\pi y/b)}{\sin(\pi x/b)\cosh(\pi y/b)+1}\right\} \\ & -\frac{2T}{\pi}\tan^{-1}\left\{\frac{\cos(\pi x/b)\sinh(\pi y/b)}{\sin(\pi x/b)\cosh(\pi y/b)-1}\right\} + 2T \end{aligned}$$

Method II:

By the method of argument principle we have in the w plane

$$\phi \;=\; 2T + \frac{1}{\pi}[(0-2T)\arg(w-1) + (T-0)\arg(w+1)]$$

which can be written as

$$\phi = \frac{T}{\pi}\tan^{-1}\left(\frac{v}{u+1}\right) - \frac{2T}{\pi}\tan^{-1}\left(\frac{v}{u-1}\right) + 2T$$

Now using the conformal transformation given by $w = \sin(\pi z/b)$, the solution in the z plane can be obtained as

$$\begin{aligned} \phi \;=\; & \frac{T}{\pi}\tan^{-1}\left\{\frac{\cos(\pi x/b)\sinh(\pi y/b)}{\sin(\pi x/b)\cosh(\pi y/b)+1}\right\} \\ & -\frac{2T}{\pi}\tan^{-1}\left\{\frac{\cos(\pi x/b)\sinh(\pi y/b)}{\sin(\pi x/b)\cosh(\pi y/b)-1}\right\} + 2T \end{aligned}$$

These two solutions are identical.

Example 2

Find the steady state temperature at any point of the region shown shaded in Fig. 7.18(a), if the temperatures are maintained as indicated.

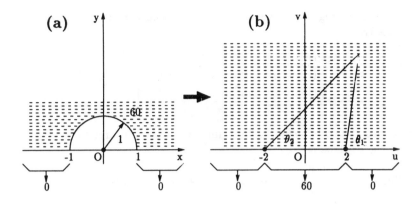

Fig. 7.18.

Solution

The shaded region of the z-plane (Fig. 7.18(a)) can be mapped onto the upper half of the w-plane (Fig. 7.18(b))by the transformation $w = z + 1/z$.

This is equivalent to $u = x + x/(x^2 + y^2)$ and $v = y - y/(x^2 + y^2)$.

The solution of the problem in the w-plane can be obtained as

$$\phi = A\theta_1 + B\theta_2 + C = A \arg(w - 2) + B \arg(w + 2) + C$$

which is the real part of the complex potential $W = C - i(A \ln(w - 2) + B \ln(w + 2))$ in which A, B and C are real constants. To determine A, B and C note that the boundary conditions on the u axis are:

$\phi = 0$	for $u > 2$	i.e. $\theta_1 = \theta_2 = 0$
$\phi = 60$	for $-2 < u < 2$	i.e. $\theta_1 = \pi, \theta_2 = 0$
$\phi = 0$	for $x < -1$	i.e. $\theta_1 = \pi = \theta_2$

These boundary conditions are obtained from the given transformation $w = z + 1/z$. Thus using these conditions, we obtain,

$$\left. \begin{array}{l} A = 60/\pi \\ B = -60/\pi \\ C = 0 \end{array} \right\}$$

Therefore, the solution to the problem in the w-plane is

$$\phi = \frac{60}{\pi} \tan^{-1} \left(\frac{v}{u - 2} \right) - \frac{60}{\pi} \tan^{-1} \left(\frac{v}{u + 2} \right)$$

Then, substituting the values of u and v, the solution to the required problem in the z-plane is

$$\phi = \frac{60}{\pi} \tan^{-1}\left\{\frac{y(x^2 + y^2 - 1)}{(x^2 + y^2 + 1)x - 2(x^2 + y^2)}\right\}$$
$$- \frac{60}{\pi} \tan^{-1}\left\{\frac{y(x^2 + y^2 - 1)}{(x^2 + y^2 + 1)x + 2(x^2 + y^2)}\right\}$$

The same result can be achieved by using the method of argument principle which simply in the w-plane is

$$T(u, v) = 0 + \frac{1}{\pi}[(60 - 0)\arg(w - 2) + (0 - 60)\arg(w + 2)]$$
$$= \frac{60}{\pi}\left[\arctan\frac{v}{u - 2} - \arctan\frac{v}{u + 2}\right]$$

Now using the conformal mapping stated above we arrive at the same result.

Exercises

1. Find the steady-state temperature distribution in the upper half of the z-plane if the portion of the x-axis between $x = -1/2$ and $x = 1/2$ is maintained at the temperature $0°C$ while the rest of the x-axis is maintained at $100°C$.

2. Work Exercise 1 if the portion of the x-axis between $x = -1/2$ and $x = 1/2$ is maintained at $100°C$ w hile the rest of the x-axis is maintained at $0°C$.

3. A semi-infinite sheet of metal whose sides coincide with the boundaries $x = -a$, $x = a$ and $y = 0$ in the z-plane has the temperatures $50°C$, $100°C$ and $0°C$ respectively. Find the steady-state temperature distribution in the metal.

4. Find the steady-state tempeature in a semi-infinite slab coinciding with the interior of a $45°$ angle standard position in the z-plane if $100°C$ along the horizontal side of the angle and $0°C$ along the other side.

5. Work Exercise 4 with the interior angle $60°C$ while the temperatures are maintained at $0°C$ and $100°C$ along the horizontal side and the other side of the angle respectively.

6. Find the steady-state temperature distribution in a circular metal disk of radius a if the upper half of the circumference of the circle is kept at temperature $0°C$ while the lower half of the circumference is kept at $100°C$.

7. Find the steady-state temperature distribution in the upper half of the z-plane exterior to the unit circle if $100°C$ is maintained along the linear portion of the boundary and $0°C$ along the circular portion of the boundary. [Hint: First show that $w = z + 1/z$ maps the portion of the upper half of the z-plane exterior to the circle $|z| = 1$ onto the entire upper half of the w-plane. Use this result.]

8. Find the steady-state temperature distribution in the first quadrant of a unit circular disk if the radial boundaries are maintained at 0°C while the curved boundary is maintained at 100°C.

9. Find the steady-state temperature distribution in the portion of the z-plane exterior to the unit circle $|z| = 1$ while the upper half of the circumference is maintained at 100°C, while the lower half is maintained at 0°C. [Hint: First show that the bilinear transformation $w = e^{i\theta}\dfrac{z+i}{z-i}$ which maps the exterior of the unit circle in the z-plane onto the upper half of the w-plane. Use this result.]

10. Work Exercise 9 if the arc of the circle in the second quadrant is maintained at 100°C while the rest of the circumference is maintained at 0°C.

7.4 Applications to electrostatic potentials

Before we demonstrate the applications to the electrostatic potentials, we first define some important quantities related to these potentials.

Coulomb's law

Let r be the distance between two point electric charges q_1 and q_2. Then by Coulomb's law, the force between them is given by

$$F = \frac{q_1 q_2}{\kappa r^2} \tag{7.47}$$

The force is called repulsion or attraction according as the charges are like (both positive or both negative) or unlike (one positive and the other negative). The constant κ is called the dielectric constant and it depends on the medium; in vacuum $\kappa = 1$, in other case $\kappa > 1$. In the following we assume $\kappa = 1$ unless otherwise specified.

Electric field intensity: electrostatic potential

Suppose we are given a charge distribution which may be continuous, discrete or a combination. This charge distribution sets up an electric field. If a unit positive charge (small enough so as not to affect the field appreciably) is placed at any point A not already occupied by a charge, the force acting on this charge is called the electric field intensity at A and is denoted by Υ. This force is derivable from a potential ϕ which is called the electrostatic potential. In symbols,

$$\Upsilon = -\text{grad } \phi = -\nabla \phi \tag{7.48}$$

If the charge distribution is two dimensional, then

$$\Upsilon = E_x + i E_y = -\frac{\partial \phi}{\partial x} - i\frac{\partial \phi}{\partial y} \tag{7.49}$$

where

$$E_x = -\partial\phi/\partial x \atop E_y = -\partial\phi/\partial y \Big\}$$

In that situation if E_t denotes the component of the electric field intensity tangential to any simple closed curve C in the z-plane, then

$$\oint_C E_t ds = \oint_C E_x dx + E_y dy = 0 \tag{7.50}$$

Gauss's theorem

Let us consider the two dimensional charge distribution. If C be any simple closed curve in the z-plane having a net charge q in its interior (actually an infinite cylinder enclosing a net charge q) and E_n is the normal component of the electric field intensity, then Gauss's theorem states that

$$\oint_C E_n ds = 4\pi q \tag{7.51}$$

If C does not include any net charge, this reduces to

$$\oint_C E_n ds = \oint_C E_x dy - E_y dx = 0 \tag{7.52}$$

It follows that in any region not occupied by a charge

$$\frac{\partial E_x}{\partial x} + \frac{\partial E_y}{\partial y} = 0 \tag{7.53}$$

which reduces to (after using (7.49))

$$\frac{\partial^2\phi}{\partial x^2} + \frac{\partial^2\phi}{\partial y^2} = 0 \tag{7.54}$$

i.e. ϕ is harmonic at all points not occupied by a charge.

The complex electrostatic potential

From the above it is evident that there must exist a harmonic function ψ conjugate to ϕ such that

$$W(z) = \phi(x,y) + i\psi(x,y) \tag{7.55}$$

is analytic in any region not occupied by charge. We call $W(z)$ the complex electrostatic potential, or, briefly, complex potential. In terms of this, (7.49) becomes

$$\Upsilon = -\frac{\partial \phi}{\partial x} - i\frac{\partial \phi}{\partial y}$$

$$= -\frac{\partial \phi}{\partial x} + i\frac{\partial \psi}{\partial x}$$

$$= -\frac{\overline{dW}}{dz} = -\bar{W}'(z) \tag{7.56}$$

The curves (cylindrical surfaces in three dimensions)

$$\left.\begin{array}{l} \phi(x,y) = \alpha \\ \psi(x,y) = \beta \end{array}\right\} \tag{7.57}$$

are called equipotential lines and flux lines respectively.

Line charges

The analogy of the above with fluid flow is quite apparent. The electric field in electrostatic problems corresponds to the velocity field in fluid flow problems, the only difference being a change of sign in the corresponding complex potentials.

The idea of sources and sinks of fluid flow have corresponding analogs for electrostatics. Thus, the complex electrostatic potential due to a line charge q per unit length at z_0 (in a vacuum) is given by

$$W(z) = -2q \ln(z - z_0) \tag{7.58}$$

and represents a source or sink according as $q < 0$ or $q > 0$. Similarly, we talk about doublets or dipoles, etc. If the medium is not a vacuum, we replace q in (7.58) by q/κ.

Conductors

If a solid is perfectly conducting, i.e. a perfect conductor, all charges are located on its surface. Thus, if we consider the surface represented by the simple closed curve C in the z-plane, the charges are in equilibrium on C and hence C is an equipotential line. An important problem is the calculation of potentials due to a set of charged cylinders. This can be accomplished by using conformal mapping.

Capacitance

Two conductors having charges of equal magnitude q but of opposite sign, have a difference of potential, say, V. The quantity C is defined by

$$q = CV \tag{7.59}$$

depends only on the geometry of the conductors and is called the capacitance. The conductors themselves form what is called a condenser or capacitor.

Example 1

(a) Find the complex potential due to a line of charge q per unit length perpendicular to the z-plane at $z = 0$.

(b) If the line is at $z = a$, what is the complex potential?

(c) Discuss the similarity with the complex potential for a line force or sink in the fluid flow.

Solution

(a) The electric field due to a line charge q per unit length is radial as shown in the following figure (Fig. 7.19).

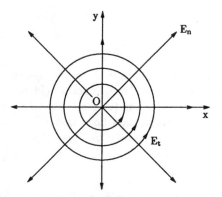

Fig. 7.19: Electric field.

The normal component of the electric vector is $E_n = E_r$ (a constant). The tangential component $E_t = 0$. If C is any cylinder of radius r, with axis at $z = 0$, then by Gauss's theorem

$$\oint_C E_n ds = \oint_C E_r ds = E_r \cdot \oint_C ds = E_r \cdot 2\pi r = 4\pi q$$

such that

$$E_r = \frac{4\pi q}{2\pi r} = \frac{2q}{r}$$

Since $E_r = -\partial\phi/\partial r$ we have after omitting the constant of integration $\phi = -2q \ln r$. This is the real part of $W(z) = -2q \ln z$ which is therefore the required complex potential.

(b) If the line charge q is at $z = a$, the complex potential is

$$W(z) = -2q \ln(z - a)$$

(c) It is to be noted that the complex potential has the same form as that for a line source of fluid if $m = -2q$. If q is a positive charge, this corresponds to a line sink.

Example 2

(a) Find the potential due to a line charge q per unit length at $z = z_0$ and a line charge $-q$ per unit length at $z = \bar{z}_0$.

(b) Show that the potential due to an infinite plane kept at zero potential (ground potential) and a line charge q per unit length parallel to this plane can be found from the result in (a).

Solution

(a) The complex potential due to two line charges is

$$W(z) = -2q\ln(z - z_0) + 2q\ln(z - \bar{z}_0)$$
$$= 2q\ln\left(\frac{z - \bar{z}_0}{z - z_0}\right)$$

Then the required potential in the real part of this

$$\phi = 2q\,Re\left[\ln\left(\frac{z - \bar{z}_0}{z - z_0}\right)\right]$$

(b) To prove this we must show that the potential

$$\phi = 2q\,Re\left[\ln\left(\frac{z - \bar{z}_0}{z - z_0}\right)\right]$$

reduces to zero on the x-axis (see Fig. 7.20(a) and 7.20(b)).

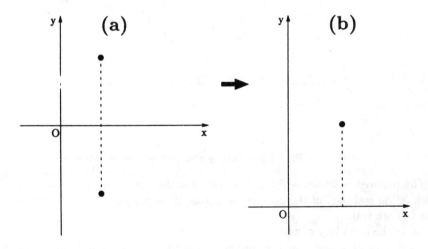

Fig. 7.20.

This follows at once from the fact that on the x-axis, $z = x$ so that

$$W = 2q \ln \left(\frac{x - \bar{z}_0}{x - z_0} \right)$$

and

$$\bar{W} = 2q \ln \left(\frac{x - z_0}{x - \bar{z}_0} \right) = -W$$

i.e. $W + \bar{W} = 0$ which means $\phi = Re\{W\} = 0$ on the x-axis. Thus, we can replace the charge q at $\bar{z}_0$ by a plane ABC at potential zero and conversely.

Example 3

(a) Find the potential at any point of the region shown in Fig. 7.21 if the potential on the axis is given by V_0 for $x > 0$ and $-V_0$ for $x < 0$.

(b) Determine the equipotential and flux lines.

Solution

(a) We must find a function, harmonic in the plane, which takes the values V_0 for $x > 0$, i.e. $\theta = 0$ and $-V_0$ for $x < 0$, i.e. $\theta = \pi$ as shown in Fig. 7.21.

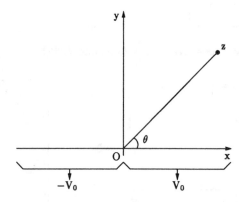

Fig. 7.21: Sketch of a half plane.

The required potential which is harmonic has the form $\phi = A\theta + B = A \arg z + B$ which is the real part of the complex potential $W = \phi + i\psi = A(-i \ln z) + B$ where A and B are real.

Using the boundary conditions

$$\text{at} \qquad \left. \begin{array}{ll} \theta = 0, & \phi = V_0 \\ \theta = \pi, & \phi = -V_0 \end{array} \right\}$$

we obtain $A = -2V_0/\pi$ and $B = V_0$ so that the required potential can be written as

$$\phi = V_0 \left(1 - \frac{2}{\pi}\theta\right) = V_0 \left(1 - \frac{2}{\pi}\tan^{-1}(y/x)\right)$$

in the upper half plane $y > 0$. The potential in the lower half plane is obtained by symmetry.

This result can also be obtained by simply using the argument principle which yields

$$\phi(x,y) = V_0 + \frac{1}{\pi}[(-V_0 - V_0)\arg(z - 0)]$$

$$= V_0 \left[1 - \frac{2}{\pi}\arctan\frac{y}{x}\right]$$

(b) The equipotential lines are given by $\phi = \alpha$, i.e. $y = mx$ where m is a constant. These are straight lines passing through the origin.

The flux lines are orthogonal projections of the lines $y = mx$. Therefore the slope of the equipotential lines is given by $y' = m$. Eliminating m we obtain $y' = y/x$ and hence the slope of the flux lines is $y' = -x/y$. This is a first order ordinary differential equation and solution of which yields

$$x^2 + y^2 = \beta^2$$

where β is an arbitrary constant and happens to be the radius of the circle. Thus the flux lines are circles of radius β with centre at the origin. The graphical representation of these orthogonal families is shown in Fig. 7.22.

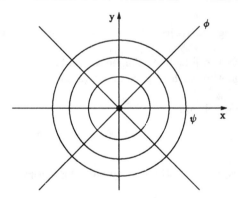

Fig. 7.22. Orthogonal trajectories of the flux lines and equipotential lines.

Example 4

The lines of electric force (flux lines) of two opposite charges of the same strength at $(-1,0)$ and $(1,0)$ can be represented by the circles $x^2 + (y - c)^2 = 1 + c^2$ passing through these points where c is an arbitrary constant. Show that the equipotential lines (orthogonal trajectories) are the circles $(x + k)^2 + y^2 = 1 + k^2$, where $c \neq k \neq 0$.

Solution

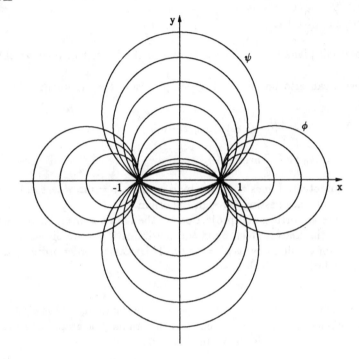

Fig. 7.23. Orthogonal trajectories of flux lines and equipotential lines.

The given equation for the flux lines can be simplified to yield $x^2 + y^2 - 2yc = 1$. Differentiating this equation with respect to x and after a little reduction yields $y' = -x/(y - c)$.

Now eliminating c we obtain

$$y' = -\frac{2xy}{y^2 - x^2 + 1}$$

Thus the differential equation of the orthogonal trajectories, i.e. equipotential lines is therefore

$$y' = \frac{y^2 - x^2 + 1}{2xy}$$

Simplifying the above equation we obtain

$$2xy\,dy - y^2\,dx = (1 - x^2)dx$$

To obtain the solution of this differential equation we separate the variables dividing it by x^2 and the resulting equation can simply be put as

$$\frac{2xy\,dy - y^2\,dx}{x^2} = \left(\frac{1}{x^2} - 1\right) dx$$

$$\left(\frac{y^2}{x}\right) = \left(\frac{1}{x^2} - 1\right) dx$$

This can be integrated very easily and thus we obtain

$$y^2 = -1 - x^2 - 2kx$$

where k is a suitable arbitrary constant. Simplifying the last equation we obtain the required result $(x + k)^2 + y^2 = k^2 - 1$.

The graphical representation of these two orthogonal families is depicted in Fig. 7.23.

Exercises

1. Two semi-infinite plane conductors making the interior of a 60° angle in the z-plane, are charged to constant potentials v_1 and v_2 respectively. Find the potential and electric field everywhere in the region.

2. Find the potential in the first quadrant of the region between the planes $y = 0$ and $x = 0$ if the potentials on the positive x and y axis are kept at constant potentials v_0 and $-v_0$ respectively.

3. Derive a formula for the potential in the region interior to an infinitely long, thin, conducting cylinder of radius a if the first quadrant is grounded and the rest of the surface is kept at v_0 potential.

4. Work Exercise 3, if the cylinder is a unit cylinder and if the potentials are zero and unity on the first quadrant and the rest of the surface respectively. Prove that the potential on the axis of the cylinder is 3/4.

5. Find a bilinear transformation which will map the exterior of the circle $|z| = b$ into the upper half of the w-plane.

6. An infinitely long, thin, conducting cylinder of radius b is split into two halves. The upper half is kept at a potential V_0 and the lower half at $-V_0$. Determine the potential distribution both inside and outside the cylinder. [Hint: Use the result of Exercise 5 to obtain the potential outside the cylinder.]

7. An infinitely long, thin, conducting circular cylinder of radius b is split into four quarters. The quarter cylinders in the second and fourth quadrants are grounded, and those in the first and third quadrants are kept at potentials V_0 and $-V_0$ respectively. Determine the potential distribution both inside and outside the cylinder.

8. Find the potential in the region $0 < r < 1$, $0 < \theta < \pi/4$ formed by the half planes $\theta = 0$ and $\theta = \pi/4$ and the portion $0 \leq \theta \leq \pi/4$ of the cylindrical surface $|z| = 1$ when the potentials 1 and 0 are on the plane boundary and on the curved boundary respectively.

Chapter 8

Analytic continuation

8.1   Introduction

We shall now study the existence of analytic continuation of a function into a domain in the complex z plane. Let us assume that we do not know the precise form of an analytic function $f(z)$ but only know that inside some circle of convergence C_0 with centre z_0, $f(z)$ is represented by a Taylor series

$$\sum_{n=0}^{\infty} a_n(z - z_0)^n \tag{8.1}$$

Choosing a point z_1 inside C_0, we can compute the values of the infinite series and all its derivatives at the point z_1 (differentiating the series term by term) and thus arrive at a new series

$$\sum_{n=0}^{\infty} b_n(z - z_1)^n \tag{8.2}$$

having circle of convergence C_1. If C_1 extends beyond C_0, then the values of this new series and its derivatives can be obtained in this extended portion and so we have achieved more information about the function $f(z)$.

In general this new series will converge in the interior of a circle C_1 which lies partly inside and partly outside C_0. When this is the case, the second series is said to be an **analytic continuation** of the first, since it provides a series representation of $f(z)$ beyond that provided by the first series (Fig. 8.1). This process can be repeated indefinitely. By, for instance, choosing a point z_2 inside C_1, we can arrive at a new series having the circle of convergence C_2 which may extend beyond C_0 and C_1. The collection of all such power series representations (all possible analytic continuations) is defined as the analytic function $f(z)$ and each power series is called an *element* of $f(z)$.

Example 1

Show that the series

$$\text{(a) } f_1(z) = \sum_{n=0}^{\infty} \frac{z^n}{2^{n+1}} \quad \text{and} \quad \text{(b) } f_2(z) = \sum_{n=0}^{\infty} \frac{(z-i)^n}{(2-i)^{n+1}}$$

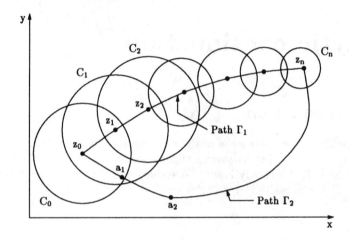

Fig. 8.1. Physical interpretation of analytic continuation into circular domains.

are analytic continuations of each other.

Solution

(a) By the ratio test, the series converges within the circle $|z| < 2$ (Fig. 8.2). In this circle the series can be summed by using the binomial expansion formula for $1/(1 - z) = 1 + z + z^2 + z^3 + z^4 + z^5 + \cdots$ for $|z| < 1$. Hence the function which represents the series is $1/(2 - z)$. If we denote the value of the series by $f(z) = 1/(2 - z)$, it is clear that this function is analytic everywhere in the complex plane except at $z = 2$.

(b) By the ratio test, the series converges for $|(z - i)/(2 - i)| < 1$, i.e. $|z - i| < \sqrt{5}$ which is a circle of radius $\sqrt{5}$ and with centre at i (see Fig. 8.2). By using the binomial expansion formula, the series can be summed and represents the function $1/(2 - z)$ which is exactly identical with the $f(z)$ obtained in (a).

Since the two power series represent the same function in the domain common to the interior of the circles $|z| = 2$ and $|z - i| = \sqrt{5}$, it follows that they are analytic continuations of each other. The function $f(z)$ is the analytic continuation of the first and also the second series into the entire complex plane except, of course, at the point $z = 2$.

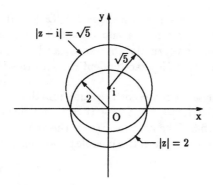

Fig. 8.2.

Example 2

Show that the series (a) $\sum_{n=0}^{\infty} z^n$ and (b) $\sum_{n=0}^{\infty} ((z+i)/i)^n$ are both analytic continuations of each other. Graph the domain of convergence for each series. Then find the function which represents all possible analytic continuations of these two series.

Solution

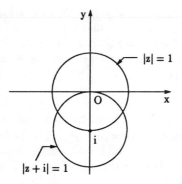

Fig. 8.3.

(a) By the ratio test, the series converges within the circle $|z| < 1$ (Fig. 8.3). In this unit circle the series can be summed by using the binomial expansion formula for $1/(1-z)=1 + z + z^2 + z^3 + z^4 + z^5 + \cdots$ for $|z| < 1$. Hence the function which represents the series is $1/(1-z)$. If we denote the value of the series by $f(z) = 1/(1-z)$, it is clear that this function is analytic everywhere in the complex plane except at $z = 1$.

(b) By the ratio test, the series converges for $|(z+i)/i| < 1$, i.e. $|z+i| < 1$ which is a circle of unit radius and centre at $-i$ (see Fig. 8.3). By using the binomial expansion formula, the series can be summed and represents the function $1/(1-z)$ which is exactly identical with the $f(z)$ obtained in (a).

Since the two power series represent the same function in the domain common to the interior of the circles $|z| = 1$ and $|z+i| = 1$, it follows that they are analytic continuations of each other. The function $f(z)$ is the analytic continuation of the first and also the second series into the entire complex plane except, of course, at the point $z = 1$.

8.2   Natural boundary of an analytic function

In performing analytic continuations we have to be careful about any singular point inside the extended domain. For instance, there cannot be any singular point (see Fig. 8.1) inside C_1 and on the boundary of C_0, since otherwise the series (8.2) would diverge at that point. It may happen sometimes that there may be an infinite number of singular points on the boundary of a circle of convergence in which case it is absolutely impossible to perform analytic continuation. In such cases the boundary of the circle is called a **natural boundary** or barrier. The function represented by a series having a natural boundary is called a **lacunary** function.

Another important question arises in the process of analytic continuation. In going from initial circle of convergence C_0 to the circle C_n (see Fig. 8.1), we have chosen the path of centres $z_0, z_1, z_2, ..., z_n$ which we may represent as Γ_1. We may arrive at C_n via the path represented by Γ_2 using the path of centres $a_0, a_1, a_2, ..., z_n$. It is absolutely remarkable that if these two paths have no singularity, we must obtain the same series representation valid inside C_n.

Example 3

Show that no analytic continuation beyond $|z| = 1$ is possible for the function represented by the series

$$1 + z + z^2 + z^4 + z^8 + \cdots \quad = 1 + \sum_0^\infty z^{2^n}$$

Solution

Let us define $f(z) = 1 + z + z^2 + z^4 + z^8 + \cdots$. Then note that

$$
\begin{aligned}
f(z) &= z + f(z^2) \\
&= z + z^2 + f(z^4) \\
&= z + z^2 + z^4 + f(z^8) \\
&= \quad ....
\end{aligned}
$$

It is to be noted first that $z = 1$ is a singular point of the given series. It is now clear from the above series representation that not only the value of z given by $z = 1$ is a singular point but also the values of z such that $z^2 = 1$, $z^4 = 1$, $z^8 = 1 \cdots$ are also singular points of $f(z)$. The distribution of this infinite number of singularities around the circumference of the circle of convergence $|z| = 1$ constitutes an impossible barrier and analytic continuation beyond $|z| = 1$ is therefore impossible. In this case the circle $|z| = 1$ is called a **natural boundary**.

Example 4

(a) Prove that the infinite series given by $\sum_{n=0}^{\infty}(-1)^n z^{n+1}$ is analytic in the domain $|z| < 1$.

(b) Find a function which represents all possible analytic continuations of the above series.

Solution

(a) By the ratio test, the series converges for $|z| < 1$, i.e., $|z| < 1$ (domain D_1), which is a unit circle and centre at the origin (see Fig. 8.4). By using the binomial expansion formula, the series can be summed and represents the function $f(z) = z/(1 + z)$.

(b) We have seen in (a) that the sum of the given series is $f(z) = z/(1 + z)$. This function is analytic at all points in the complex domain except at $z = -1$ (domain D_2), at which the function goes to infinity. Since $f(z)$ is equal to the series inside $|z| = 1$, hence $f(z)$ is the analytic continuation of the given series into the complex domain. Thus this is the required function.

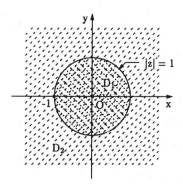

Fig. 8.4. Domain D_1 is a subset of domain D_2 except at $z = -1$.

8.3 Generalization of analytic continuation

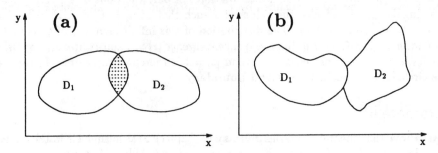

Fig. 8.5. Sketch for analytic continuation into arbitrary domain.

Let us suppose that $f_1(z)$ is an analytic function of z defined in a domain D_1 (Fig. 8.5(a)). If we can find a function $f_2(z)$ which is analytic in a domain D_2 such that $f_1(z) = f_2(z)$ in a domain common to D_1 and D_2 in that situation we say that $f_2(z)$ is analytic continuation of $f_1(z)$ into the domain D_2. This means that there is a function $f(z)$ analytic in the combined domain of D_1 and D_2 such that $f(z) = f_1(z)$ in D_1 and $f(z) = f_2(z)$ in D_2. In actual practice it suffices for D_1 and D_2 to have only a small arc as shown in Fig 8.5(b).

By analytic continuation to domain D_3, D_4 etc., we can extend the original domain of definition to other parts of the complex plane. The functions $f_1(z)$, $f_2(z)$, $f_3(z)$, $\cdots$ defined in D_1, D_2, D_3, $\cdots$ respectively are called the **elements** of the function $f(z)$. It is sometimes impossible to extend a function analytically beyond the boundary of a domain. We then call the boundary a **natural boundary**.

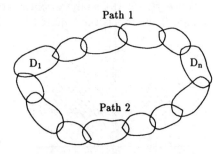

Fig. 8.6. Uniqueness theorem for analytic function.

If a function $f_1(z)$ defined in D_1 is continued analytically to domain D_n along two paths (see Fig. 8.6), then the two analytic continuations will be identical provided there is no singularity between the paths. This is the uniqueness theorem for analytic continuation.

We have already seen how functions represented by the power series may be continued analytically. In this section we consider how functions with other representations such as integrals may be continued analytically.

Example 5

(a) Prove that the integral defined by $\int_0^\infty te^{-zt}dt$ is analytic at all points z for which $Re(z) > 0$.

(b) Find a function which represents all possible analytic continuations of this integral.

Solution

(a) The given integral is recognized as the Laplace transform of t valid for the domain $Re(z) > 0$ and can be evaluated pretty quickly as follows:

$$\mathcal{L}\{t\} = f_1(z) = \int_0^\infty te^{-zt}dt = \frac{1}{z^2}$$

(b) For $Re(z) > 0$, the integral has the value $f_2(z) = 1/z^2$ which is analytic at all points in the complex domain except at $z = 0$. Figure 8.7 depicts the domains of $Re(z) > 0$ which is D_1 and $z \neq 0$ which is D_2. Since $f_2(z) = f_1(z)$ for D_1, we see that $f_2(z) = 1/z^2$ is the required analytic continuation of $f_1(z)$ into the domain D_2.

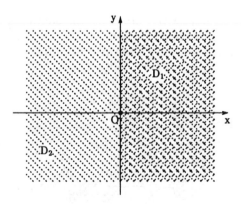

Fig. 8.7. Domain D_1 is a subset of domain D_2 except at z=0.

Example 6

(a) Show that the integral given by $\int_0^\infty (1+t)e^{-zt}dt$ converges only if $Re(z) > 0$.

(b) Find a function which represents all possible analytic continuations of this integral.

Solution

(a) The given integral is recognized as the Laplace transform of $1 + t$ valid for the domain $Re(z) > 0$ and can be evaluated pretty quickly as follows:

$$\mathcal{L}\{1 + t\} = f_1(z) = \int_0^\infty (1 + t)e^{-zt}dt = \frac{1 + z}{z^2}$$

(b) For $Re(z) > 0$, the integral has the value $f_2(z) = (1 + z)/z^2$ which is analytic at all points in the complex domain except at $z = 0$. Figure 8.7 depicts the domains of $Re(z) > 0$ which is D_1 and $z \neq 0$ which is D_2. Since $f_2(z) = f_1(z)$ for D_1, we see that $f_2(z) = (1 + z)/z^2$ is the required analytic continuation of $f_1(z)$ into the domain D_2.

Example 7

(a) Find the domain of convergence of the integral $\int_0^\infty e^{-(1+z)^2}dt$ and graph this domain.

(b) Find the value of the analytic continuation of the integral corresponding to $z = 1 - 2i$.

Solution

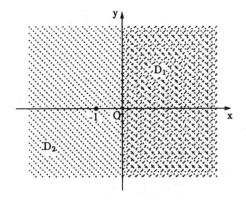

Fig. 8.8.

(a) The given integral can be evaluated in the domain $Re(z) > 0$ at once as follows:

$$f_1(z) = \int_0^\infty e^{-(1+z)^2 t}dt = \frac{1}{(1 + z)^2}$$

(b) For $Re(z) > 0$, the integral has the value $f_2(z) = 1/(1+z)^2$ which is analytic at all points in the complex domain except at $z = -1$. Figure 8.8 depicts the domains of $Re(z) > 0$ which is D_1 and $z \neq -1$ which is D_2. Since $f_2(z) = f_1(z)$ for D_1, we see that $f_2(z) = 1/(1+z)^2$ is the required analytic continuation of $f_1(z)$ into the domain D_2.

The value of the analytic continuation of $f_1(z)$ corresponding to $z = 2 - 4i$ is given by

$$f_2(2 - 4i) = \frac{1}{(1 + 2 - 4i)^2} = \frac{-7 + 24i}{625}$$

8.4 Some important theorems

Theorem 8.1

If $f(z)$ is analytic in a domain D and that $f(z) = 0$ at all points on an arc AB inside the domain then $f(z) = 0$ throughout D (Fig. 8.9).

Proof

Choose any point, say z_0, on arc AB. Then in some circle of convergence C with centre at z_0 which extends at least to the boundary of D where a singularity may exist, $f(z)$ can be expanded by Taylor's series

$$f(z) = f(z_0) + f'(z_0)(z - z_0) + f''(z_0)\frac{(z - z_0)^2}{2!} + \cdots$$

But by hypothesis $f(z_0) = f'(z_0) = f''(z_0) = \cdots = 0$, and hence $f(z) = 0$ inside C.

By choosing another arc inside C, we can continue the process. In this way we can prove that $f(z) = 0$ throughout D. Hence the required proof.

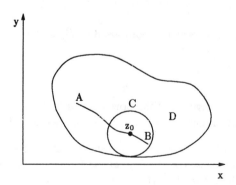

Fig. 8.9.

Example 8

Given that the identity $\sin^2 z + \cos^2 z = 1$ holds for all real values of z then prove that it also holds for all complex values of z.

Solution

Let $f(z) = \sin^2 z + \cos^2 z - 1$ and let D be a domain of the z plane containing a portion of the x axis.

Since $\sin z$ and $\cos z$ are analytic in D, it follows that $f(z)$ is analytic in D. Also it is clear that $f(z) = 0$ on the x axis. Hence by Theorem 8.1, $f(z) = 0$ identically in D, which shows that

$$\sin^2 z + \cos^2 z = 1$$

for all z in D. Since D is arbitrary, we obtain the required result.

The following theorems are cited here without proof. Interested readers are referred to standard textbooks, including Churchill and Brown (1990).

Theorem 8.2

If $f_1(z)$ and $f_2(z)$ are analytic within a domain D and suppose that on an arbitrary arc in D, $f_1(z) = f_2(z)$, then prove that $f_1(z) = f_2(z)$ in D.

Theorem 8.3

If $f_1(z)$ is analytic in domain D_1 and on an arbitrary boundary and if we can find a function $f_2(z)$ analytic in domain D_2 and on the same boundary such that $f_1(z) = f_2(z)$ on the boundary, then prove that the function

$$\begin{aligned} f(z) &= f_1(z) \text{ for } z \text{ in } D_1 \\ &= f_2(z) \text{ for } z \text{ in } D_2 \end{aligned}$$

is analytic in the domain D which is composed of D_1 and D_2.

Exercises

1. If the radius of convergence of the power series $\sum_{n=0}^{\infty} a_n(z - z_0)^n$ is r, prove that the radius of convergence of the termwise derivative is also r.

2. If $\sum_{n=0}^{\infty} a_n(z - z_0)^n$ converges to $f(z)$, show that in the interior of the circle of convergence, the derivative of $f(z)$ can be found by termwise differentiation of the series and the integral of $f(z)$ can be found by termwise integration of the series.

3. Find the Maclaurin expansion of

$$f(z) = \int_C \frac{dt}{t - z} \quad |z| \neq 1$$

if C is the circle $|t| = 1$. Show that this expansion converges for all values of z but represents $f(z)$ only in the interior of C. [Hint: Use Cauchy's integral formula.]

4. Show that the function $f_2(z) = 1/(1 + z^2)$ for $z \neq \pm i$ is the analytic continuation of the function $f_1(z) = \sum_{n=0}^{\infty} (-1)^n z^{2n}$ for $|z| < 1$ into the domain consisting of all points in the z plane except $z = \pm i$.

5. Show that the function $f_2(z) = 1/z^2$ for $z \neq 0$ is the analytic continuation of the function $f_1(z) = \sum_{n=0}^{\infty} (n + 1)(z + 1)^n$ for $|z + 1| < 1$ into the domain consisting of all points in the z plane except $z = 0$.

6. Find the analytic continuation of the function $f(z) = \int_0^{\infty} t^n e^{-zt} dt$ where $Re(z) > 0$ into the domain consisting of all points in the z plane except the origin.

7. Show that the function $z/(1 + z^2)$ is the analytic continuation of the function $f(z) = \int_0^{\infty} e^{-zt} \cos t \, dt$ where $Re(z) > 0$ into the domain consisting of all points in the z plane except $z = \pm i$.

Bibliography

- Ahlfors, L.V.(1966). *Complex Analysis*, 2nd ed., McGraw-Hill Book Company, Inc., New York.

- Bronwell, A.(1953). *Advanced Mathematics in Physics and Engineering*, McGraw-Hill, New York, pp. 386-413.

- Chesnut, H. and Meyer, R.W.(1951). *Servo-mechanisms and Regulating System Design*, John Wiley, New York.

- Churchill, R.V.(1972). *Operational Mathematics*, 3rd ed., McGraw-Hill, New York.

- Churchill, R.V.(1963). *Fourier Series and Boundary Value Problems*, 2nd ed., McGraw-Hill, New York.

- Churchill, R.V. and Brown, J.W.(1990). *Complex Variables and Applications*, McGraw-Hill, New York.

- Copson, E.T.(1957). *Theory of Functions of a Complex Variable*, Oxford University Press, London.

- Curle, N. and Davies, H.(1968). *Modern Fluid Dynamics*, Vol. 1. Van Nostrand, London.

- Dettman, J.W.(1965). *Applied Complex Variables*, Macmillan Company, New York.

- Dirac, P.M.A.(1935). *The Principles of Quantum Mechanics*, Oxford University Press, Oxford.

- Forsyth, A.R.(1918). *Theory of Functions of a Complex Variable*, Cambridge University Press, London.

- Guilleman, E.A.(1953). *The Mathematics of Circuit Analysis*, John Wiley & Sons, Inc., New York.

- Jeans, J.H.(1925). *Mathematical Theory of Electricity and Magnetism*, Cambridge University Press, Cambridge.

- Kaplan, W.(1973). *Advanced Calculus*, 2nd ed., Addison-Wesley Publishing Company, Inc., Reading, Mass.

- Kellogg, O.D.(1953). *Foundations of Potential Theory*, Dover Publications, New York.

- Lamb, H.(1945). *Hydrodynamics*, Dover Publications, New York.

- Lebedev, N.N.(1965). *Special Functions, and their Applications*, Prentice-Hall, Inc., Englewood Cliffs, N.J.

- Lighthill, J.(1986). *An Informal Introduction to Theoretical Fluid Mechanics*, Oxford University Press, Oxford.

- Love, A.E.H.(1944). *Elasticity*, Dover Publications, New York.

- McLachlan, N.W.(1963). *Complex Variable Theory and Transform Calculus*, Cambridge University Press, Cambridge.

- Milne-Thomson, L.M.(1955). *Theoretical Hydrodynamics*, Macmillan & Co., Ltd., London.

- Rahman, M.(1991). *Applied Differential Equations for Scientists and Engineers Vol. 1, Ordinary Differential Equations; Vol. 2, Partial Differential Equations*, Computational Mechanics Publications, Southampton UK and Boston USA.

- Rahman, M.(1994). *Water Waves: Relating Modern Theory to Advanced Engineering Applications*, Oxford University Press, Oxford.

- Sneedon, I.N.(1951). *Fourier Transforms*, McGraw-Hill, New York.

- Spiegel, M.R.(1964). *Complex Variables*, McGraw-Hill, New York.

- Thaler, G.J. and Brown, R.G.(1960). *Analysis and Design of Feedback Control Systems*, McGraw-Hill, New York.

- Titchmarsh, E.C.(1939). *Theory of Functions*, Oxford University Press, London.

- Whittaker, E.T., and Watson, G.N.(1950). *Modern Analysis*, Cambridge University Press, London.

- Wylie, C.R., and Barrett, L.C.(1982). *Advanced Engineering Mathematics*, McGraw-Hill, New York.

Answers to odd-numbered exercises

Chapter 1

Section 1.2

1. $1 + 7i$

3. $19 + 22i$

5. $\frac{2}{5} + \frac{2}{5}i$

7. $\frac{1}{2} + \frac{\sqrt{3}}{2}i; \frac{1}{2} - \frac{\sqrt{3}}{2}i$

9. $\frac{\sqrt{3}}{2} + \frac{1}{2}i; -\frac{\sqrt{3}}{2} + \frac{1}{2}i; -i$

11. $(1,2); \left\{(-1 + \sqrt{17})/2, (9 + \sqrt{17})/16\right\}; \left\{(-1 - \sqrt{17})/2, (9 - \sqrt{17})/16\right\}$

13. $1 - iz$

17. $r^{\frac{1}{5}} \left[\cos\{(\theta + 2k\pi)/5\} + i \sin\{(\theta + 2k\pi)/5\}\right]$ where $r = \sqrt{11}, \theta = \tan^{-1}(\sqrt{2}/3)$, $k = 0, 1, 2, 3, 4$

21. $2^{\frac{1}{6}} \left[\cos(\pi/12) + i \sin(\pi/12)\right]; 2^{\frac{1}{6}} \left[\cos(3\pi/4) + i \sin(3\pi/4)\right];$ $2^{\frac{1}{6}} \left[\cos(17\pi/12) + i \sin(17\pi/12)\right]$

23. $\cos 3\theta = 4\cos^3 \theta - 3\cos \theta; \sin 3\theta = 3 \sin \theta - 4 \sin^3 \theta$

Section 1.3

1. (a) $z \neq 1$; (b) $z \neq \pm i$; (d) $z \neq 0$

3. $10 + 3i$

5. $\bar{z}^2 + 2iz$

11. (a) discontinuous at $z = \pm 1$; (b) discontinuous at $z = i$; (c) discontinuous at $z = -i$; (d) discontinuous at $z = i$

13. differentiable at $z = 0$ only

Section 1.4

11. (a) $v(x, y) = e^x \sin y + c$
    (b) $v = x^2 - y^2 + 6y + c$
    (c) $v = x^2 - y^2 - 2x + c$
    (d) $v = 3xy^2 - 2x + c$
    (e) $v = -\sinh x \sin y + c$

Section 1.5

7. (a) $(2n+1)\pi i, n = 0, \pm 1, \pm 2, \cdots$
   (b) $\frac{1}{2}\ln 7 + (2n\pi + \tan^{-1}\frac{\sqrt{3}}{2})i, n = 0, \pm 1, \pm 2, \cdots$
   (c) $2 + (2n + \frac{1}{2})\pi i, n = 0, \pm 1, \pm 2, \cdots$
   (d) $\ln(2 \pm \sqrt{5}) + 2n\pi i, n = 0, \pm 1, \pm 2, \cdots$
   (e) $(2n+1)\pi i, n = 0, \pm 1, \pm 2, \cdots$

11. (a) $(2n+1)(\pi/2) + i\ln(\cosh 2 \pm \sinh 2)$ if n is even;
    $(2n+1)\frac{\pi}{2} + i\ln(-\cosh 2 \pm \sinh 2)$ if n is odd (b) $n\pi + i\ln(8 \pm \sqrt{15})$ if n is even;
    $n\pi + i\ln(-8 \pm \sqrt{15})$ if n is odd.

15. $\cosh x \cos y = c_1; \sinh x \sin y = c_2$

17. (a) $e^{-\pi/2}$; (b) $e^{-\pi/2}[\cos(\ln 2) + i\sin(\ln 2)]$;
    (c) $e^{-3\pi \tan^{-1}(-\sqrt{e})}[\cos(3\pi \ln(\frac{e\sqrt{1+e}}{2})) + i\sin(3\pi \ln(\frac{e\sqrt{1+e}}{2}))]$

Chapter 2

Section 2.2 and 2.3

1. (a) $-1 + i$; (b) $-\frac{3}{2} + \frac{5}{2}i$

3. (a) $\frac{5}{3}(2 - i)$; (b) $\frac{1}{3}(14 + 11i)$; (c) $\frac{1}{3}(2 - 13i)$

5. $2\pi i \sin 4, 0$

7. $2\pi i, 0$

9. (a) except $z = 0$; (b) except $z = (2n+1)\frac{1}{2}, n = 0, \pm 1, \cdots$; (c) all z;
   (e) except $z = -2 \pm 3i$;(f) all z

Chapter 3

Section 3.1

1. $|z| < 1$

3. $|z - 1| < 1$

5. $|(z+2)/(z-2)| < 2$

9. $x^4 + y^4 - 2x^2 + 2y^2 + 2x^2y^2 < 0$

11. $x > 0$

13. $x < 0$

Section 3.2

1. $-[1 + 2\sum_{n=1}^{\infty} z^n]$ about $z = 0, |z| < 1$; $-\sum_{n=1}^{\infty}(z+1)^n$ about $z = -1, |z+1| < 1$

3. $-\sum_{n=1}^{\infty}(-1)^n\{(z - \frac{\pi}{2})^{2n-1}/(2n - 1)!\}$, for all z

5. $(\pi/4) + (1/2)\sum_{n=1}^{\infty}(-1)^{n-1}\{(z - 1)^n/n!\}$

Section 3.3

1. $\sum_{n=0}^{\infty} 2^n/z^{n+1}$, for $|z| > 2$; $-\sum_{n=0}^{\infty} z^n/2^{n+1}$, for $|z| < 2$

3. $(7/4) + (3/2)(z - 1)^{-1} + \sum_{n=1}^{\infty} 7(z - 1)^n/2^{n+2}$ for $|z - 1| < 2$;
$2/(z - 1) - 1/(z - 1)^2 + 2/(z - 1)^3 - 4/(z - 1)^4 + \cdots$ for $|z - 1| > 2$

5. (a) 0; (b) $2\pi i$; (c) $2\pi i$; (d) 0; (e) 0

7. (a) $\sum_{n=1}^{\infty}(-1)^{n+1}(n + 1)(z - 1)^{n-1}$, for $|z - 1| < 1$

   (b) $1/(z - 1) + \sum_{n=1}^{\infty}(-1)^n n/(z - 1)^{n+2}$, for $|z - 1| > 1$

Chapter 4

Section 4.1 and 4.2

1. (a) $\{z/(1 - iz)\}(1 - iz - z^2 + iz^3 + z^4 - \cdots)$, $z = i$ isolated singularity
   (b) $\sum_{n=1}^{\infty}(-1)^{n+1}z^{n+1}$, $z = -1$ isolated singularity
   (c) $\sum_{n=-1}^{\infty} z^{-n}/(n + 1)!$, $z = 0$ essential singularity
   (d) $\sum_{n=0}^{\infty} z^{2n}/(2n + 1)!$, $z = 0$ removable singularity
   (e) $\sum_{n=0}^{\infty}(-1)^n z^{2n-1}/(2n)!$; $z = 0$ isolated singularity
   (f) $\sum_{n=0}^{\infty} z^{-2n+1}/n!$, $z = 0$ essential singularity

3. (a) 0, (b) $-\frac{1}{6}$, (c) $-\frac{1}{2}$

5. $\frac{1}{2}$ at $z = 1$, $(15 + 49i)/12$ at $z = 3i$, and $(15 - 49i)/12$ at $z = -3i$

7. 0

9. (a) $2\pi i$, (b) $-2\pi i$, (c) 0, (d) $2\pi i$

Section 4.3

1. $\pi/2$

3. $\pi/(2\sqrt{2})$

5. $(\pi/2a)e^{-ab}$

7. $\pi/\{2(a + b)\}$

9. $\{\pi/2(a^2 - b^2)\} ((e^{-b\alpha}/b) - (e^{-a\alpha}/a))$

11. $\pi(e^{-b\alpha} - e^{-a\alpha})/(a^2 - b^2)$

13. $(\pi/\sin p\pi)(b^{p-1} - a^{p-1})/(a - b)$

15. $(\pi a^{p-3}/3 \sin p\pi)(1 + 2\cos(2\pi p/3))$

17. $2\pi(2x^2 - 1)/(x^2 - 1)$

19. $2\pi a/(a^2 - b^2)^{3/2}$

21. $\{(2\pi a)/(b^2\sqrt{a^2 - b^2}f)\}(a - \sqrt{a^2 - b^2})$

Chapter 5

Section 5.1

1. $b_n = 0, a_0 = 16/3, a_n = (-1)^{n+1}(16/n^2\pi^2)$

3. (a) $c_n = (-1)^n\{(1 + in)/(1 + n^2)\pi\}\sinh \pi, n = 0, \pm1, \pm2, \cdots$

7. $b_n = 0, a_0 = 4, a_n = -16/n^2\pi^2$, for n odd otherwise 0

Section 5.2

1. (a) $\sin(b\sigma)/b\sigma$; (b) 1

3. (a) $-(4/\sigma^3)(\sigma \cos \sigma - \sin \sigma)$; (b) $-\pi/4$

Section 5.4

1. $S\{f'\} = -nC\{f\}; C\{f'\} = (-1)^n f(\pi) - f(0) + nS\{f\}$

3. $(1/2) - e^{-t} + (e^{-t}/2) - ((1/2) - e^{-(t-1)} + (e^{-(t-1)}/2)) u(t - 1)$

5. $y(t) = te^{-t}$; answer to the first one

7. (a) $(2/s) + 5$; (b) $2s/(s^2 + 4)$; (c) $1 - e^{-2s}$; (d) $1 - 5s$; (e) $2 - s + s^2 e^{-s}$

Section 5.6

1. $(1/2a^3)(\sin at - at \cos at)$

3. $(e^{-t}/4)(2t \cos 2t + \sin 2t)$

5. $\{1/(2\sqrt{\pi})\} \int_0^t (1/\sqrt{\tau}) \sinh 2(t - \tau)d\tau$

7. $u(t)$

Section 5.7

1. $(e^{-at}/b)\sin bt$

3. te^{-3t}

5. $(e^{-2t}/3)\sin 3t$

7. $(e^{t/2}/3)\left[\sqrt{3}\sin(\sqrt{3}t/2) + \cos(\sqrt{3}t/2)\right] - (e^{-t}/3)$

9. $(e^{-t}/16)[\sin 2t - 2t\cos 2t]$

11. $(t/b) + (2/\pi)\sum_{n=1}^{\infty}\{(-1)^n/n\}\sin(n\pi t/b)$

13. $(e^{-at}/\cosh ab) + 4\sum_{n=0}^{\infty}\dfrac{(-1)^{n+1}\left[(2n+1)\pi\cos\frac{(2n+1)\pi t}{2b} - 2ab\sin\frac{(2n+1)\pi t}{2b}\right]}{(2n+1)^2\pi^2 + 4a^2b^2}$

Chapter 6

Section 6.2

3. $|w| = a^2$

7. $u = x^3 - 3xy^2, v = 3x^2y - y^3; 27u^2 = (v+1)(v-8)^2$

9. $u = 1 - y/(x^2 + y^2), v = -x/(x^2 + y^2)$

11. $u^2 + v^2 = e^{2\alpha}, v = u\tan\beta$

Section 6.3

1. $v = 3x^2y - y^3, w = z^3$

7. (a) $80/3$

Section 6.4

1. $w = (6iz + 2)/(z + 3i)$

3. $w = i/z$

7. (a) $1 + \sqrt{2}, 1 - \sqrt{2}$, (b) $(1 + i\sqrt{9})/2, (1 - i\sqrt{9})/2$

9. a, b, c, d real, and $ad - bc < 0$

11. $w = (z^4 - i)/(z^4 + i)$

21. $50 + (50/\pi)\tan^{-1}\left(4y/(x^2 + y^2 - 4)\right)$

23. $50\left[1 - \tan^{-1}(v/u)\right]$

Chapter 7

Section 7.2

1. (a) $y = $ constant, $x = $ constant; (b) $y = k/(x+2)$, $(x+2)^2/k - y^2/k = 1$
   (c) $e^{-x\ln 3}\sin(y\ln 3) = k$, $e^{-x\ln 3}\cos(y\ln 3) = k'$
   (d) $\cos x \sinh y = k$, $\sin x \cosh y = k'$

3. $\phi = m\ln(r_1 r_2/r)$, $\psi = m(\theta_1 + \theta_2 - \theta)$ where $r_1 = \sqrt{(x+a)^2 + y^2}$,
   $r_2 = \sqrt{(x-a)^2 + y^2}$, $r = \sqrt{x^2 + y^2}$ and $\theta_1 = \tan^{-1}(y/(x+a))$,
   $\theta_2 = \tan^{-1}(y/(x-a))$, $\theta = \tan^{-1}(y/x)$

5. $W = m\ln\{(z-a)/(z+a)\}$, $\phi = m\ln(r_2/r_1)$, $\psi = m(\theta_2 - \theta_1)$
   where $r_1 = \sqrt{(x+a)^2 + y^2}$, $r_2 = \sqrt{(x-a)^2 + y^2}$, $\theta_1 = \tan^{-1}(y/(x+a))$,
   $\theta_2 = \tan^{-1}(y/(x-a))$

7. $\phi = x^2 - y^2$, $\psi = 2xy$; (a) $\vec{q} = (4, -2)$, (b) $\vec{q} = (-4, -2)$

9. $\phi = \frac{1}{2}\cos^{-1}(x^2 + y^2)$, $\psi = \frac{1}{2}\cosh^{-1}(x^2 + y^2)$

11. $W = -(iK/2\pi)\ln(z+a) + (iK/2\pi)\ln(z-a)$, $\phi = (K/2\pi)[\tan^{-1}(y/(x+a))$
    $\qquad - \tan^{-1}(y/(x-a))]$, $\psi = (K/2\pi)\ln(r_2/r_1)$

    where $r_1 = \sqrt{(x+a)^2 + y^2}$, $r_2 = \sqrt{(x-a)^2 + y^2}$

13. $150 - \{25\ln(x^2 + y^2)\}/\ln 2$

Section 7.3

1. $(100/\pi)\tan^{-1}\{4y/(4x^2 + 4y^2 - 1)\} + 100$

3. $(50/\pi)\left[2\tan^{-1}\{(\cos\frac{\pi x}{2a}\sinh\frac{\pi y}{2a})/(\sin\frac{\pi x}{2a}\cosh\frac{\pi y}{2a} - 1)\}\right.$

   $\qquad\qquad \left. - \tan^{-1}\{(\cos\frac{\pi x}{2a}\sinh\frac{\pi y}{2a})/(\sin\frac{\pi x}{2a}\cosh\frac{\pi y}{2a} + 1)\}\right]$

5. $(100/\pi)\tan^{-1}\{(3x^2 y - y^3)/(x^3 - 3xy^2)\}$

7. $(100/\pi)\tan^{-1}\frac{4(x^2+y^2)y(x^2+y^2-1)}{4(x^2+y^2)^2 - x^2(x^2+y^2+1)^2 - y^2(x^2+y^2-1)^2} + 100$

9. $-(100/\pi)\tan^{-1}\{(x^2 + y^2 - 1)/2y\} + 100$

Section 7.4

1. $\{(v_2 - v_1)/\pi\}\tan^{-1}\{(3x^2 y - y^3)/(x^3 - 3xy^2)\} + v_1$

3. $(v_0/\pi)[\tan^{-1}\{(1 - x^2 - y^2)/2y\} - \tan^{-1}\{(1 - x^2 - y^2)/(2y - (x+1)^2 - y^2)\}]$
   $\qquad + v_0$

5. $w = i(z + b)/(z - b)$

7. $(v_0/\pi)\,[\tan^{-1}\{(x^2 + y^2 - b^2)/(2by + b[(x - b)^2 + y^2])\}$
$\qquad - \tan^{-1}\{(x^2 + y^2 - b^2)/2by\}$
$\qquad - \tan^{-1}\{(x^2 + y^2 - b^2)/(2by - b[(x - b)^2 + y^2])\}\,] + v_0$

Chapter 8

Section 8.4

3. $f(z) = 2\pi i$

Index